About Science

About Science

Barry Barnes

Basil Blackwell

First published 1985

Basil Blackwell Ltd
108 Cowley Road, Oxford OX4 1JF, UK

Basil Blackwell Inc.
432 Park Avenue South, Suite 1505,
New York, NY 10016, USA

British Library Cataloguing in Publication Data

Barnes, Barry
About science.
1. Science——Social aspects
I. Title
306′.45 Q175.5

ISBN 0-631-14157-X
ISBN 0-631-14158-8 Pbk

Library of Congress Cataloging in Publication Data

Barnes, Barry.
About science.
Bibliography: p.
Includes index.
1. Science. 2. Science——Social aspects.
3. Research.
I. Title.
Q160.2.B37 1985 500 85-6011

ISBN 0-631-14157-X
ISBN 0-631-14158-8 (pbk.)

Typeset by Freeman Graphic, Tonbridge, Kent
Printed and bound in Great Britain at
The Camelot Press Ltd, Southampton

Contents

Preface ix

Acknowledgements xiii

1 *The Rise of Science*

GROWTH 1
THE CONTEXT OF GROWTH 13
MODERN SCIENCE 20
SCIENCE, TECHNOLOGY AND ARMAMENTS 29

2 *Science for its own Sake*

RESEARCH 37
WHO COUNTS? 49
WHAT COUNTS? 59

3 *Authority*

AN EXPERIMENT 72
INTERPRETATION 75
THE AUTHORITY OF SCIENCE 79

4 *Expertise in Society*

SCIENTISM 90
TECHNOCRACY 98
ANOTHER VIEW 104

5 *Thoughts on the Future*

POSSIBILITIES 113
IMPOSSIBILITIES 123
DANGERS 132

Appendix 145

Notes and References 153

Further Reading 159

Index 161

Preface

Modern science, by which I mean scientific research and all the various tasks and undertakings associated with it and required by it, is at once one of the most practically important and one of the most intrinsically interesting of all current human activities. Not surprisingly, therefore, it has attracted considerable curiosity, and has been studied in considerable depth and from many points of view. Quite a detailed picture can now be drawn of it, using the work of historians, psychologists, sociologists and anthropologists, economists and political scientists, and many others whose concern is to describe and explain human behaviour in its many forms. And the picture is full of interest and fascination. Natural scientists themselves, however, and science students, are generally too involved in the study of nature to be able to turn more than a small part of their attention directly upon science itself. Although many of them would wish to do so, they lack the time to read into the relevant materials. They cannot hope to search out and assimilate all the different available perspectives on science, all the relevant ways of thinking, all the forms of curiosity and kinds of question to ask, all the tools and procedures brought to bear.

In this book I have tried to make the situation just a little bit less difficult by bringing some of this material together and indicating a little of its value and significance in the course of a general discussion of natural science. Needless to say I can only begin to discuss the various themes and topics introduced in the text, and to hint at their full interest and importance: I cannot hope to present them in adequate detail or at a proper level of precision. What I hope is that, whilst avoiding the complexities and fine points which try the patience and spoil the enjoyment of the independent reader, I can nonetheless do some kind of justice to the materials I discuss, and perhaps encourage some further reading.

The book is primarily written for people actively involved in natural science, especially for those studying it, or thinking of studying it, in some form of higher education. But there is no reason why it should not be taken up by the general reader. It is a book *about* science, not a book of science; its concern is with science as an activity, with the way that science is ordered and organized, and particularly with the relationship of science to the rest of society. Anyone with an interest in these matters should find the text accessible enough.

Although my own scientific training and continuing involvement with natural science had a great deal to do with my decision to write this book, I do not presently work as a scientist and I have not attempted to present problems and situations from a scientist's viewpoint. How scientists themselves typically see their work and its significance is, of course, a topic of immense importance. But it is readily ascertained by reading the words of natural scientists themselves, in their essays and autobiographies, and the appropriate parts of their journals and periodicals. I have tried here to expand and complement the inside view, not to reinforce it, to go beyond what is typically encountered in the course of an education in the natural sciences, and point out, as it were, some additional resources for the mind. And I have tried to indicate some of the profound and deep-seated problems which have fallen upon us as a consequence of the rise of science, as well as the enormous range of benefits of which scientists and students of science will already be well aware.

Since it discusses the natural sciences as a whole, my book must necessarily concentrate on general themes and issues, and leave others to convey a sense of the achievements of particular scientific fields and the experience of working within them. I can offer but a distant vision of science and its social context. Such a distant vision may however have its merits. One can miss phenomena by using too high-powered a microscope objective, as well as by staying at too low a power: it may be depth of field which matters, not resolving power. Similarly, with science, it is as well to step back and review the scene, to correct as well as to confirm close up perceptions. It is very important, in connection with this, to bear in mind the enormously fragmented and specialized nature of our current natural science, and the small proportion of the whole routinely perceptible to many of those hard at work on the research front within it. As Robert Oppenhauer pointed out, many years ago now, much of the current public ignorance of science is shared by scientists themselves:[1]

> Today, it is not only that our kings do not know mathematics, but our philosophers do not know mathematics and – to go a step further – our mathematicians do not know mathematics. Each of them knows a branch

> of the subject and they listen to each other with a fraternal and honest respect; and here and there you find a knitting together of the different fields of mathematical specialisation . . . We so refine what we think, we so change the meaning of words, we build up so distinctive a tradition, that scientific knowledge today is not an enrichment of the general culture. It is, on the contrary, the possession of countless highly specialised communities who love it, would like to share it, and who make some efforts to communicate it; but it is not part of the common human understanding . . . We have in common the simple ways in which we have learned to live and talk and work together. Out of this have grown the specialised disciplines like the fingers of the hand, united in origin but no longer in contact.

I shall have quite a lot to say about specialization and its consequences in the course of the book. It has profound effects not just upon the ordering of life and our relationships with other people, but on the way that we think, and actually upon the way that we know. Since we live today in a highly specialized society, with an intense division of intellectual labour in which natural science is very heavily involved, both its advantages and the problems it presses upon us are of enormous practical interest.

I have myself experienced some of the problems of specialization, in a very trivial form, in the writing of this book. Needless to say, being an academic, I am myself a specialist. There are, I admit, one or two topics in this very wide-ranging book which I actually do know something about: the problem with them is that I know that I have treated them too simply and missed out important and relevant aspects of them. But most of the material to be discussed I simply do not know enough about. Nor have I been able to train up properly in the half a dozen or so special fields in which ideally I ought to have been competent. I have simply had to take the plunge and discuss what I felt to be important issues, on the basis of what specialists in these fields would rightly call inadequate knowledge. Be warned: there are bound to be more than the usual number of errors in this book.

What I must hope is that any errors or lapses in the presentation of materials normally dealt with by various kinds of specialist will be tolerable in a book designed more to raise questions and offer tools of thought than to convey specific facts or findings.

Acknowledgements

Kim Pickin first suggested that this book needed to be written and persuaded me to set aside the necessary time: I am grateful to her and other staff at Basil Blackwell, both for this initial stimulus and for continuing patience and encouragement thereafter. Many of the ideas in the book, and something of the style in which they are discussed, must derive from my own work environment, although in saying this I do not wish to deny my own full responsibility for the text, together with all the errors and infelicities which it contains. What I do wish is to indicate the extent of my indebtedness to all the staff at the Science Studies Unit at Edinburgh University, and to the many natural science students who have attended my courses there over what is now the better part of two decades. In some way, I am not sure how, this book is a product of that long experience. Finally I wish to thank Donald MacKenzie and Carole Tansley for their particular assistance, deeply appreciated in both cases.

Barry Barnes
Edinburgh

1

The Rise of Science

GROWTH

The way of life established and taken for granted today in societies like our own is one which has emerged only very recently and which is altogether without historical precedent. Fundamentally, our current state is the outcome of industrialisation: we are amongst the first generations ever to possess the vast material resources of a developed industrial society, to experience the demands it makes in terms of social organization and social relationships, to plan and act in ways conditioned by the environment it has created. But we can also think of the same development as a transformation at the level of culture, thought and sensibility: we have become a secular society based upon technical impersonal knowledge, a society which grants to scientists and scientific knowledge the place which our predecessors allowed to priests and religious doctrines.

The social change and the cultural change really do seem to run together, and to constitute different aspects of a single historical development. The large, clearly visible events generally referred to as 'the rise of industry' and 'the rise of science' both took place in the course of the last three and a half centuries. Science and industry seem to have arisen in parallel, and to have made their immense impact jointly over this very short period. I should say their impact so far. It would be quite wrong to think of our modern society as a stable condition at which we have arrived. For good or ill, we are still travelling, whilst the majority of the population of the world seem intent upon arriving at what currently is our point of departure.

Since science is our subject, let me turn at once to the rise of science and try to convey a sense of what sort and size of development it was. Think first of a typical academic scientist, working in the way we are familiar with today, in a university perhaps, carrying out the normal

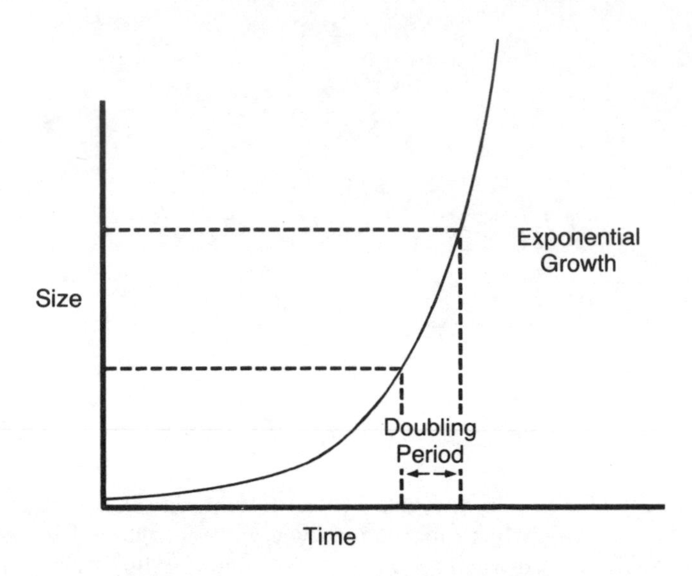

Figure 1.1 The exponential growth curve

duties of teaching and research. Such a scientist may readily find the time to train or supervise two or more graduate students at any one time. Let two students be typical, and assume that they graduate after three years' work. The scientist will thus be training two fully competent potential colleagues every three years, ten every 15 years. Naturally there will be some 'wastage' of these trained people, some tendency for them to move out of science. And of those who remain, and work themselves as scientists, only a proportion will be able to engage in graduate teaching, in propagating the scientific species in the way that they themselves were propagated. But even if only two out of every ten were to do so, our original typical scientist would still have produced two fertile successors in a 15 year period, successors who would themselves each be equally capable of producing two such successors in another 15 years. Even if our typical scientist were to teach no more at the end of 15 years, he would have replaced himself with two equally fertile offspring. If he really were typical, then science would double in size every 15 years.

Clearly, as far as the logistics of the situation are concerned, this represents a relaxed and easily achieved rate of reproduction. If the outside society declined to interfere, and allowed scientists to draw in whatever in the way of support and nourishment they needed, then they could double their numbers every 15 years or so, effortlessly, without placing any stress upon their internal procedures and internal resources. This is an exponential rate of growth; it produces a growth curve of ever increasing steepness, and hence over a long period can produce some mind-boggling changes, as can be seen in figure 1.1. A population which doubles in size every 15 years increases by about 100 times over a century. Over three centuries growth would be a million-fold: one scientist in 1660 would become a million scientists by 1960. Exponential growth can bring forth a great deal from small beginnings: the unconstrained increase of animal populations is described by an exponential curve, as every rabbit-breeder knows.

The actual quantitative growth of science has in fact been studied in considerable detail. The first relevant materials were gathered together in the early 1960s by the historian of science, D. J. de Solla Price, and are still available in his celebrated book *Little Science: Big Science.*[1] Price systematically counted all the quantitative indicators he could find of the growth of science: numbers of scientists, of journals, of articles published in journals, and so on. The different figures indicated roughly the same pattern for the entire period he was able to study. For two hundred years or more science had grown exponentially, doubling in size every 15 years or so. What we saw to be possible on the basis of the unconstrained activity of scientists, appears to have been actual in the history of science itself.

Needless to say, Prices' exponential curves provide only a very general, low-resolution image of the overall pattern of growth of science. Individual sciences or branches of science have varied widely in their growth characteristics. So has the rise of science in particular countries and particular institutions. And even at their best, the data are rather more bumpy and irregular than the smooth curves drawn through them might suggest. The low resolution image is nonetheless very useful. For example, the manifest steepness of the growth curve brings home very vividly indeed what Price calls the 'immediacy' of science. At any point in its period of exponential growth, the working population of scientists actually constitutes most of the scientists who have ever lived. In a certain sense, 'typical' science is always the current science: science swamps its own history. I myself find this a particular salutary thought. Although I am not a professional historian, I make considerable use of the literature of the history of science, and I like to believe that there is a very great deal of current relevance to be learned from history. Yet one has only to go back a century or so into the history of science to find oneself far into the tail of the growth curve, studying an activity but one-hundredth the size of its present-day equivalent. There must be an analogy of some kind between the place of, say, Charles Darwin in the history of science, and that of Shakespeare or the first Elizabeth in the mainstream of our history. By the same analogy, the great controversy between Newton and Leibniz would correspond to that between Harold and William the Conqueror. Clearly, great care is necessary in using the history of even quite recent science in attempting to understand that of the present day, although that is not to say that no such attempt should be made: people may not have changed all that much since the Norman Conquest.

In making this analogy I have assumed that science has grown much more rapidly than most of the rest of society, so that 15 years for the history of science is rather like 50 years for the history of society generally. The assumption is a reasonable one: a 50 year doubling period, for example, will describe the general rate of increase in population from the seventeenth century, and of the size of the labour force. Productivity has increased much more rapidly, but still at a substantially slower rate than science. The growth of science, in other words, has consistently outstripped that of the resources and infrastructure needed to support and sustain it. For a long period, this was of no real importance: no doubt even the British economy could readily absorb the impact as the number of its scientists doubled from ten to 20 to 40. But in the end science was bound to grow to the extent that it became a significant consumer of social and economic resources. In the mid-1960s, soon after Price's book was published, there were of the order of 150,000 scientifically and technically qualified people engaged in research and development in Britain, about half of whom were fully qualified scientists or engineers. Perhaps this is not an unduly large proportion of a labour force of 20 millions, or even of the most talented people in such a labour force, but it is a large proportion on which to base continuing exponential growth. After another century of such growth, given our current population trends, the entire labour force would have been conscripted to science.

The point is only reinforced if the cost rather than the size of science is considered. The cost of science has actually tended to increase historically with a doubling period of about ten years or so. Perhaps this has had something to do with the ever-increasing use of costly technology by scientists, culminating in the 'big science' of the post-war period with its accelerators, radio-telescopes, immense computers, and so forth. In any event, this rate of growth of costs implies an increase of a thousandfold over a century. But expenditure on science in the 1960s was already very high: typically in the developed countries research and development accounted for over 2 per cent of Gross National Product at this time.

Clearly, there was no way in which science could hope to maintain an exponential rate of expansion: it was pressing hard against external

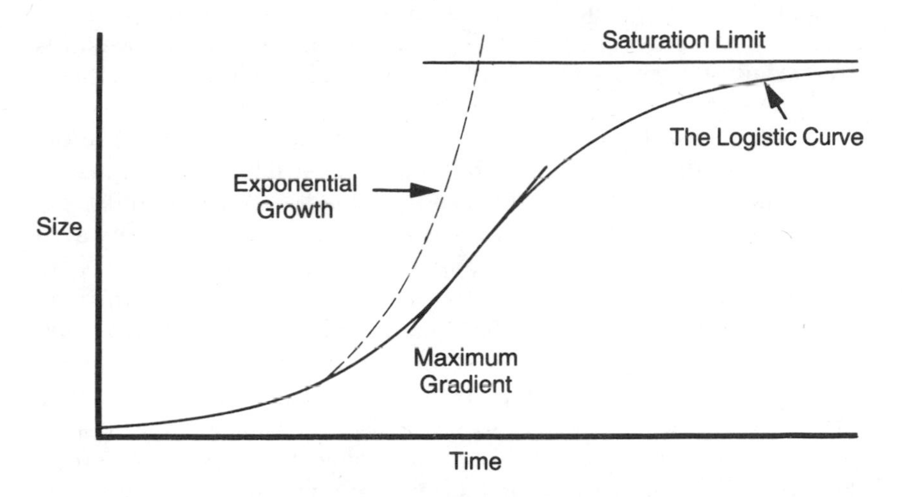

Figure 1.2 The logistic curve

constraints. The internal tendency to exponential growth would now be dampened by the external tendency to begrudge the necessary resources. The growth curve, which had hitherto thrust more and more steeply upwards, would now be wrenched round and out, more and more toward the horizontal. Price himself therefore suggested that an S-shaped 'logistic' curve was the one with which to describe scientific growth, a curve little different from an exponential in its early stages but then increasingly falling away to lower and lower growth rates, as shown in figure 1.2. It may be that Price was unwise always to try to describe the vagaries of social and economic change with precise mathematical curves, but basically his prediction was correct: the rate of scientific growth has fallen off very markedly since the early 1960s; its subsequent development, whatever the details, has certainly been closer to the logistic than to the exponential curve.

As a consequence, the general perception of science and the subjective experience of being a scientist have both begun to change over the last 25 years or so. Throughout its history science had always been perceived as a rapidly expanding activity burgeoning forth from within, and having an ever-increasing effect on the outside society through its discoveries and innovations. But now that it has arrived in a big way it is perceived as a cost as well as a benefit, as something affected by society as well as something which affects it. It is now becoming increasingly accepted that levels of scientific activity should be determined by political processes, and made to reflect so-called 'economic realities'.

And political and economic considerations are more than ever influencing the direction of scientific work as well as its amount. Scientists are having to wean themselves off a supply of resources that hitherto had grown as fast or even faster than their aspirations, and adjust to a lower rate of growth. They are having to think more and more seriously of how to justify their research to those key outsiders who ultimately control the purse-strings, which means that considerations of utility, and short-term utility at that, are receiving greater and greater priority even in universities and other institutions of what used to be called 'pure science'. Moreover, the actual process of slowing down growth is in some ways more painful than a finally adjusted state of no growth at all. The adjustment invites all sorts of temporary disruptions: job opportunities may grow much less rapidly than the hitherto rapidly expanding supply of qualified graduates, so massively diminishing career opportunities; funds for new developments may almost disappear altogether, so that earlier open ended long-term commitments can be met; new generations of technology and equipment may not be made available. Subjectively, reductions in the rate of growth of something may be perceived and experienced as a contraction. This kind of perception is now a familiar one in many scientific fields and disciplines, and will continue to be so for some time to come if Price's predictions about the future development of science prove to be justified. According to his conjectures, science, at least in the developed countries, is now just over half-way through a difficult transition period, likely to last for half a century or a little more. During the transition it will adjust from the condition of exponential expansion to one much closer to a stable state, wherein it grows at a rate little different from that of the economy generally.

I shall come back to the present state of science and the extent to which it is now bound up with our political and economic institutions. But let me leave this for the moment, and move back toward the other end of the growth curve. From which point should we date the rise of science, and why? Needless to say, there is no question of following the exponential curve backwards until we arrive at the first scientist. The curve has value only as a very rough and ready image of changes in the size of science of an order of magnitude or so, and even then it has value only for more recent periods of scientific growth. As we press backwards into history the size of science ceases to be comfortably larger than the kinds of errors involved in constructing the curve. And there is another, deeper, problem. It becomes more and more difficult to apply the terms 'science' and 'scientist' without stretching their standard meaning and doing violence to many of their accepted connotations. As we go to the eighteenth century, and then to the seventeenth and sixteenth, seeking an answer to the questions, 'When did science begin?' and 'Where are the first scientists?', history forces us to realize that the answer very much depends on what we wish these terms to mean. Copernicus, Galileo, Descartes, even Isaac Newton himself, all systematically differed from modern scientists, and indeed from each other, in their beliefs and practices, and in the basis of their so-called 'scientific' judgement. We have to take a view on how important these various differences were. In this respect, the problem of identifying the first scientist is just like that of identifying the first human being or the first rabbit. Biologists rightly refuse to accept problems formulated in this way. Instead, they ask how a population becomes recognizably a human population, or a rabbit population, over time. This is just how we should re-think the problem of the origins of science. We should look out not for an event, or a date, but for a drawn-out *process* wherein the activities of groupings of people gradually become more and more recognizable as scientific activities. And we should recognize that with gradual processes of this kind beginnings and endings are very much a matter of judgement.

To talk sensibly of the origins of science it is necessary to forget about numbers and curves, and to revert to common sense and qualitative history. And even then we must expect no single 'correct' answer. Many historians of scientific thought and historians of ideas speak of the 'Scientific Revolution' which occurred between about 1540 and 1700, from the initial reception of the Copernican system of astronomy to the culminating achievement of the new philosophy it inspired in the work of Isaac Newton. This is the period wherein the historian finds so many of the discoveries and reorientations in thinking which made possible the emergence of our present-day common sense view of the world. It includes any number of specific achievements in astronomy, mechanics, optics, anatomy, natural history, chemistry, and many other areas. And it encompasses a profound transformation of thought involving the demise of the man-centred, teleological cosmology of Aristotle and the Aristotelian scholars, and its replacement with what fundamentally is an impersonal and mechanical world picture. It is a period too which is rich in discussions of what we would now call 'scientific method'. The importance of our knowledge of physical nature was given a new emphasis at this time, and the various means of obtaining that knowledge were evaluated and compared: the role of observation and experiment was examined, and the need for hypothesis, for quantification and mathematization, and for the use of reliable processes of reasoning.

Is it justified to speak of the Scientific Revolution of the sixteenth and seventeenth centuries, and to mark the rise of science from the beginning of that period? I think so. It is true that practically every 'scientific idea' which had currency in this period can be shown to have had analogues in earlier times, in the mechanics of some of the scholars of the medieval period for example, or even in the literature of Ancient Greece. But the important point is that during the period of the Scientific Revolution these ideas became more and more widely accepted amongst literate people, so that finally they were securely entrenched and indeed dominant: the national scientific societies of England, France and Germany (Brandenberg) were all founded at the end of this period, between about 1660 and 1700. Moreover, the general ideas embodied in the science of this period (in the 'new philosophy' as it was then called) were ideas which established themselves in many other contexts (law, religion), and the ideas displaced likewise tended to be displaced everywhere. The seventeenth century saw the sudden and steep decline of belief in astrology and in witchcraft, for example, both being forms of knowledge which interpret natural phenomena as intimately bound up with the practical and moral concerns of men.

The rise of science in the seventeenth century seems to have been part of a major shift of thought and sensibility involving a marked decline of anthropocentrism, anthropomorphism and teleology. Man (anthropos) was no longer found in position at the centre of the universe (anthropocentrism), but was exiled to an unremarkable, unprivileged location. The explicit attribution of man's characteristics to nature (anthropomorphism) became problematic and in need of justification, whereas previously it had been commonplace and taken for granted to an extent difficult to credit today. The explanation of natural processes or natural entities in terms of their supposed purpose (teleogy), and in particular in terms of their supposed purpose of *benefiting man,* was no longer accepted. All in all, man lost his primacy in the overall scheme of things. This shift was not, of course, a complete one. Scientists and philosophers today continue to oppose anthropocentrism, anthropomorphism and teleology. But the shift was a substantial and systematic one, and its effects have endured. The seventeenth century probably deserves its status as a watershed in the history of thought and ideas.

There is, however, more to science than thought and ideas. Primarily, it is an activity. And today it is a routinely-established, securely-financed activity, an activity carried out by trained and qualified professionals. Here is the basis for an alternative, equally valid, way of thinking of the rise of science – in terms of its emergence and growth as a professional occupation. This was not something which occurred in the seventeenth century; it is necessary to move forward many years, well into the nineteenth century, to observe this development getting under way. The very term 'scientist' is a useful marker here. It seems first to have been used in 1833, when William Whewell employed it at a meeting of the British Association for the Advancement of Science to describe those there assembled. Very rapidly the term came to denote those professionally engaged in a branch of science, and for this very reason several of the finest English 'scientists' of the day disliked the term, and preferred to continue to be known as 'philosophers'. But the term 'scientist' stuck, as more and more men of science accepted an image of themselves as professionals. It is interesting to note, however, that even today the English term 'scientist' is narrower in its meaning and implications than its analogues in European languages, reflecting, perhaps, a peculiarly restricted and hard-boiled conception of scientific knowledge and scientific activity established in the English-speaking world. As compared with Continental societies, we place the scientist further from wisdom, learning and insight, and closer to mere technique; and we make a much stronger distinction than do they between 'scientists' and 'intellectuals'.

What was involved in the professionalization of science in the nineteenth century? Above all, it was a matter of job creation. In the seventeenth and most of the eighteenth centuries, for all the quality of their scientific achievements, there were scarcely any paid scientific posts. Those in England could be counted on the fingers of one hand; in France there were rather more, but no great number. Science was an amateur activity, a diversion for those with the necessary attributes of wealth and leisure: initially, in the seventeenth century, it was dominated by gentry and aristocrats, but by the late eighteenth century it had moved substantially down-market and had become predominantly middle class, reflecting the increase in numbers and resources, throughout Europe, of merchants and bankers, civil servants and bureaucrats, lawyers and other professional people. And no doubt this move down-market increased the pressure for the professionalization of science. In any event, jobs for scientists made an appearance in substantial numbers early in the nineteenth century and continued to increase on a prodigious scale as time passed. It was in the educational system that the greater proportion of posts was established, notably in the *Écoles* of post-revolutionary France and a little later in the universities of the German States. It is probably fair to say that it was the concern of governments in these countries to establish comprehensive systems of state education which provided the conditions in which the scientific career could be created.

Needless to say the foundation of professional science as we now know it was far more than a state inspired job creation scheme. Most of the institutional arrangements discussed in the next chapter were developed in this period, and an elaborate supporting infrastructure for science came into existence. For the first time, systematic training in the various fields of science became routinely available, training which could be based upon actual practice in a laboratory. And formal qualifications became associated with the various levels of training, and specific career opportunities with the qualifications. Research was supported alongside teaching, and eminent scientists could look forward to the control of their own laboratory, or even their own research institute, and to the assistance of skilled and competent technicians. This is indeed one of the most notable and significant of the nineteenth-century innovations: permanently established laboratories, the power-houses of modern science, were wellnigh unknown before this time. Finally, as the century wore on, more and more professional scientific associations came into existence, along with professional journals designed to communicate the research being carried out in the rapidly growing and rapidly fragmenting scientific community. The various scientific disciplines and specialties proliferated at a remarkable rate, and every discipline and increasingly every specialty needed a journal. Referring back to Price, whose results pose fewer problems of interpretation in this period when science was both large and well defined, we find that the number of journals and the number of journal articles grew exponentially in line with other indicators of the size of science.

Naturally, the professionalization of science, and its establishment as a securely financed occupation, proceeded differently in different countries and any general sketch must necessarily gloss over these differences and give something of a false picture. But on the other hand there was a tendency for successful institutional arrangements to develop in one country and then to be imitated elsewhere, so that a certain overall uniformity did develop. The organization of German academic science in particular was widely admired and widely used as a model; large numbers of foreign scientists trained in Germany and transplanted its system when they returned to found laboratories in their own countries. Many of the innovations described in the previous paragraph were first developed and proved within the German universities. Indeed it was in them that the stereotype of the pure, academic scientist, engaged both in the teaching of students and the performance and publication of research, first became clearly defined.

The professionalization of science was a process of truly profound significance. Its importance, its continuing, far-reaching implications, are difficult to exaggerate. One immediate consequence was, of course, that with systematic training, reliable communications, effective quality control, and technical resources, scientific research became far more efficient. But this gain in efficiency, important as it was, is actually very far from being the most significant aspect of the changes which had occurred. With the advent of the professional scientist a *social role* had been established which was specifically required and expected to develop and modify existing knowledge, and a *social institution,* science, had been ordered and arranged around this objective. As far as I know, this is quite without historical precedent. Other societies have had roles and institutions devoted to the preservation and transmission of knowledge; indeed all societies have such institutions and science is one of the institutions to perform this task in our own society. But to *extend* and *modify* knowledge, *as a matter of routine,* as a matter of the explicit standard practice of a specific occupation, that, I think, is unique. With the institutionalization of science in the nineteenth century a great engine of change was built right into the social fabric itself.

Why is it important whether or not an activity is routinized, made a part of standard social practice, 'institutionalized' as I said above? The main reason is that the processes whereby activities become institutionalized, and those whereby they cease to be institutionalized, are very different, and although the former may be difficult enough to implement, the latter are nearly always very much more so. Once some activity or practice is generally perceived as a routine and continuing part of a society, i.e. as an institution, people plan on its continued existence, and having laid their plans they stand to lose if that continued existence is not forthcoming. By their planning, people acquire a vested interest in the continuation of the institution, and hence, without any particular regard to rights and wrongs, become its supporters. Everyone who makes the effort and adjustment necessary to become a scientist bets upon the continuation of science and acquires a vested interest in it. The same is true of law and lawyers, medicine and doctors, pits and coal-miners, and so on. The insiders of any institution always have a vested interest in its continuation. But outsiders may bank upon its continuation also – by relying on the goods or services or personnel it produces, for example, or just by taking time and trouble to adjust to it. And thus additional vested interests and consequent support are created. One way or another institutions become coupled to and sustained by interests. Think of driving on the left, spelling, the two-party system, the school holidays, the schools themselves, the farm-price support systen, VHS. None of these practices is beyond abolition, but all of them have become routinized, counted upon and planned around, and henceforth linked to a wide range of interests and decidedly difficult to shift. They have all, in various ways, become locked into the institutional structure of society generally.

This is what happened with science in the course of the nineteenth century. It was locked into the institutional structure. And it has become more and more securely locked in since that point, to the extent that it is now entirely utopian to contemplate an unlocking. We are now obliged to live as no other society has, with a marvellously potent knowledge changing device more or less permanently implanted as part of our system of institutions.

With hindsight, is it not clear that this implantation has been a good thing? Has not the obligation thus forced upon us proved a positive pleasure to carry out? For myself, I would have no hesitation in answering 'yes' to both questions: yes, so far. But precisely because science is now an institution of our society, its benefits and advantages need thinking about with particular care, and in a special way.

It is well known that conservative politicians are often averse to the creation of special-purpose institutions or organizations for the solution of particular social problems. Why is this? It is, they say, because there is no way of ensuring that institutions will continue to serve the purposes for which they are initially created, and forebear from seeking to further other purposes. It is the tragedy of institutions that they take on, as it were, a life of their own. Once conjured into existence as routine and organization, they tend to drift out of contact with their initial ends, and can pose a continuing problem of control for the remaining components of society. Institutions are not permanently connected to the particular human ends they serve at any time. The connection is *contingent* not *necessary,* variable not fixed. The threat that their relationship with human ends might change is ever present from the time that the institutions are created. And the mere possibility of this occurring can be, for many people, something which quite outweighs any immediate benefit which the creation of an institution or an organization could provide.

Consider an army. An army may be created for defence against an external power. But very rapidly the routine and organization that is the army may be directed into securing the interests of its membership or of the powerful amongst its membership. Or cliques in the wider society may be able to use the army to crush their private enemies or impose a dictatorial regime. Either way, a society may be pillaged or repressed by the army initially created to defend it. The creation of an army must be thought of as the creation of a potential – a general resource, a versatile tool. The initial creation is invariably bound up with some specific conception of its use or application. But the continuing application of the tool in just that way can never be taken for granted. It is a continuing problem from the point that the army is brought into existence. For this very reason the avoidance of the creation of a standing army has been a recurrent theme in the history of English politics, and the threat it might pose to the good order of the kingdom has often been set on a par with that posed by external enemies. Indeed, even today, developed industrial and democratic societies have to worry about their armies.

Needless to say, there is no comparison to be made between science and an army. Nowhere in its history has science turned upon the hand that feeds it in the way that armies so frequently have done. But it is important to remember that science is an institution, just as an army is an institution. It is an institution the enormous potency of which has so far mainly been directed at what people consider to be worthwhile ends. But there is no necessary connection of the institution and these ends. We ourselves have to take care that the connection between science and human ends is as we would have it to be. And, however long the connection remains as we would have it, we must nonetheless never allow ourselves to forget its contingency and its insecurity. The enduring legacy of the nineteenth century is the permanent problem of the use and control of institutionalized science.

THE CONTEXT OF GROWTH

The rise of science has proceeded in parallel with industrialization and the associated transformation in our general way of life. This is surely no coincidence. It cannot be that on the one hand we were creating financial and commercial institutions of extraordinary sophistication, moving to an urban-based existence, and eventually developing productive resources on an unprecedented scale, and that on the other hand, quite independently, we were having some good ideas about nature and the laws underlying physical events and processes. The rise of science is an integral part of the emergence of modern industrial society: the most powerful productive system ever known and the most powerful system of instrumental knowledge ever known did not coincide just by chance. Yet, when matters are examined in proper detail it proves extraordinarily difficult to produce a satisfactory general account of what the connections were.

The seventeenth century, as we have seen, encompassed a remarkable transformation in our conception of nature, which cleared the ground for the growth of scientific activity and the eventual establishment of science as an occupation. But it was also in this century that the ground was cleared for the growth of industrial activity. The industrial revolution itself, of course, had yet to come, but it was at this time that so many of the ideas, attitudes and practices which later facilitated it first came into prominence: individualism, the concept of a free market in goods, the skeleton of a satisfactory banking and credit system, the legal reinforcement of financial transactions, and so on. These were the innovations of an increasingly important commercial and financial middle class. The growing importance of this class and its ideas and practices is referred to by some writers as 'the rise of mercantile capitalism', or simply as 'the rise of capitalism'.

Naturally, writers have attempted to connect the rise of science with the rise of capitalism, and to explain the former as the consequence of the latter. But there is still no generally agreed account of the connection.[2] Some writers see the rise of the new science as a response to new needs. It is said, for example, that capitalism created new economic and technological needs, and that science grew in response to these needs. Others regard not new needs but new ideas, images and ways of thinking as of primary importance. They suggest that in a capitalist society where individualistic, impersonal and mechanistic modes of thought are generally important, they will be taken up and used to interpret nature, thus giving rise to a scientific world view. Yet another view refers neither to the new ideas associated with capitalism nor to the new needs it created, but emphasizes a specific social change which capitalism engendered, and which brought two separate existing streams of knowledge and skill into contact. It is claimed that in the early capitalist society the important social barrier between scholars and craftsmen was broken down, and the old distinction between rationally based and empirically based knowledge eroded away. In the pre-capitalist society, there had been a massive social divide between practitioners of the 'liberal' and the 'mechanical' arts, just one aspect of an intense prejudice on the part of philosophers and scholars against those who worked with their hands and relied upon merely technical knowledge. But in the cities of early capitalist Europe the divide weakened, and there were practical and economic incentives to a fusion of the hitherto separate forms of knowledge. The fusion, in this view, is what gave rise to science as we now understand it, a unique combination of abstract theoretical and mathematical kinds of thinking with empirical observation and technical skill.

It is easy to understand why writers should wish to relate the Scientific Revolution and the initial rise of capitalism. The two developments run in parallel. And it is particularly striking how both developments appear alike to start in southern Europe, notably Italy, and how the centre of gravity of both slowly moves northward so that by the end of the seventeenth century the most advanced forms of capitalist society are located in the Dutch Republic and in the City of London, and it is likewise in England and Holland that science is thriving most vigorously. But although the connection is undeniably an immensely suggestive one, none of the above accounts of its basis is generally accepted, and even the existence of a causal connection of some kind between science and capitalism is disputed. I have alluded to a few of the better known attempts to account for the Scientific Revolution merely to convey a sense of the difficulty of the task. I do not intend to advocate any particular point of view here: I shall wait along with everyone else until the historians sort out the situation further.

What, however, of the later growth of paid professional science, and the flow of large amounts of money and material resources to the support of scientific research?[3] Cannot this aspect of the rise of science convincingly be connected to the parallel growth of industry and industrial society? In this case there is indeed something much closer to a general agreement that the connection does exist. But as before, it has proved peculiarly difficult to discover just what the connection is.

The history of the industrial revolution in Britain and Europe is an enormous topic, and our current knowledge still does not enable us to say with any confidence why it occurred. But it does seem to be generally accepted that industrialization was not the product of scientific advance. Over the whole period of the industrial revolution there are, of course, any number of interesting links between scientific and industrial change. But for the most part industrialization seems to have got under way on the basis of fairly simple mechanical innovations inspired by ingenuity and experience rather than scientific knowledge. This was certainly the case in the cotton textile industry which was at the forefront of the take-off into industrialization in Britain. And even where technical knowledge was important for industrial innovation and change, that knowledge was not usually scientific or science-based. There was a strong division between science and the knowledge of engineering and the practical arts throughout the eighteenth and nineteenth centuries, and it was the latter knowledge which played a role when bodies of technical knowledge had any role to play, at least until the process of industrialization was very well advanced. It is very important to be aware of this today, when the economic effects of science are taken for granted and the phrase 'science and technology' is often used as if it refers to one single thing.

If science did not produce the rise of industry was it then the effect of that rise? There is indeed a great deal which suggests that the massive growth of scientific activities and institutions over the last two centuries has been underpinned by the continuing process of industrialization. But the relationship has not been a simple one. Certainly it will not do to claim that professional science thrived and prospered purely as a response to a demand for useful knowledge. Science had gone a long way before it actually reached the stage of being useful, and hence of being able to command support because it was useful. The industrial research laboratory, for example, seems to mark a comparatively late recognition by industrialists of the value of science: systematic research had grown first in universities and educational institutions, notably in Germany, and the scientists produced by those institutions convinced industrialists of the potential utility of their fields. In the late nineteenth century, for example, in the dyestuffs industry and the metals industries, it was not so much that there was a demand from industry which called forth scientists, as that scientists arrived and created a demand for themselves. And the initial supply of scientists seems originally to have resulted more from a felt need for *pure* science and science *education* than for *applied* science and scientific *training*.

One suggestion which might help to account for this is that science was initially favoured as an appropriate form of knowledge and culture, relevant to a great range of the concerns and values of the industrializing society. In particular, science could serve as a vehicle of cultural and symbolic expression for the now rapidly expanding commercial and industrial middle classes, and could be used by them as a means of justifying themselves and their way of life. Science offered a full and comprehensive account of the world which could stand as a viable alternative, even as a formidable challenge, to the religious world-view established amongst the old landed classes of society. And although science did not at this stage in its development have great practical utility, it was a form of knowledge which promised utility, which was attractive to people who valued utility, which fitted congenially into the utilitarian approach to life of the new classes. Science, it might be said, was their style, rather as the Bible, and Greek and Latin literature, were the style of the old landed gentry and aristocracy whom they were rapidly supplanting as the dominant social and political power during the course of the nineteenth century.

It does seem reasonably well established that as the commercial and industrial middle classes increased in numbers and significance they created their own distinctive ways of living (legendary for sobriety, respectability, and cold calculation), and evolved and supported distinctive ideals and doctrines. They are known to have held the natural sciences in higher esteem than theology, classics and the other forms of traditional learning, and they made a major contribution to what was eventually an irresistible pressure to establish the sciences as recognized forms of education within the universities and similar institutions. Science was an important part of the foundations of their alternative culture. Just as the classics were an appropriate basis for the education and hence the ways of life of the old landed gentry and aristocracy, so, in just the same way, was science appropriate for the new middle classes: it was an alternative form of 'polite learning'. Just as the old aristocracy and gentry could reflect upon, legitimate and celebrate their way of life by drawing upon their classical education, so the new classes could do the same for their way of life by drawing upon the concepts, themes and imagery provided by their scientific education. The role of science was not primarily to provide specific skills, but to be the cultural and intellectual basis for a form of life which could stand as a developed alternative to that of the old landed classes. Scientific education stood to an education in the classics as the new urban, industry based form of life stood to the old, rural, agriculture based form.

If this guess is correct it explains why there was so strong a movement for science *education* rather than scientific *training*. Stereotypically, a training transmits instrumental skills of no intrinsic value. It is a mere means to an end, something provided to serve ends beyond and outside itself. As a training, science is strictly on a par with dentistry, or cabinet-making, or even less elevated forms of skilled labour: it is pure instrumentality devoid of intrinsic value, justifiable only in terms of what specific objectives it can be made to serve. But, again stereotypically, education is concerned with intrinsic values: it transmits knowledge and competence worth while in itself, and produces inherently desirable changes in those undergoing it, improved powers of inference, for example, or heightened sensibility. Thus, whereas scientifically well-trained people may do valuable service in a good society, scientifically well-educated people may *be* a good society.

Clearly, for science to serve as the cultural basis for a whole way of living it had to be seen as the product of education. It could be no mere means to an end, but had to be conceived of in terms of inherent goods. It had to be thought of not as mere technique but as a repository of truth and a paradigm of sound inference, so that to learn science was to become well informed and rational. Conceived of in this way a scientific education was a good in itself, in the same way as the classically based education to which it stood as an alternative. Instead of learning based on tradition it offered knowledge based on experience; instead of refinement and cultivation it induced rationality and objectivity; even the practical utility of scientific knowledge was presented as an intrinsic good, a feature that any form of knowledge simply ought to display. Thus, although the goods embodied in a scientific education were different, it could, in the same way as classical education, be justified as the foundation of a good society. And a scientifically educated society probably did stand for many in the new middle classes as the apotheosis of their ideals, just as a classically educated society stood similarly for many of the old landed classes.

Needless to say, as the years have gone by, the potential utility of the sciences has become actual, and the economic application of scientific procedure and scientific knowledge has increased to the extent that science is sometimes thought of simply as a part of the economy. Utilitarian considerations of various kinds are now, beyond doubt, the major underpinning for the support of science. But even today, when there is intense concern with the level of qualified 'scientific manpower' and all the pressure is to make science more and more practically relevant, we do not see science purely in terms of instrumentality and use. It is recognized that there is a basic distinction between, say, physics and dentistry, that the one is in some way, however difficult to define, on a different plane to the other.

We recognize that sciences such as physics receive support as inherently desirable activities as well as instrumentally useful ones. Support on utilitarian and on non-utilitarian grounds run together, with their relative importance varying from time to time, but with neither actually disappearing altogether. When utilitarian concerns are predominant the tendency is to talk of the need for 'trained manpower' and for more 'relevant' science. When a larger vision of science's significance obtains the talk tends to be of 'education'.

Even though the utilitarian significance of science has increased out of all recognition throughout the present century, the large conception of science as the basis of culture continues to be vigorously expressed. It is often said that a genuinely modern society, dependent upon the power of current technology and industrial organization, cannot simply rely upon its possession of technical scientific skills. Such a society, it is said, must be thoroughly at home with science; scientific culture must thoroughly permeate it; scientific concepts and images must not remain the restricted possessions of specialists and experts. Otherwise people will feel alienated from and hostile to the science and technology upon which they depend, and out of control of their own destinies.

One of the most interesting expressions of this viewpoint to appear in the post-war period is found in the writing of C. P. Snow, particularly in his essay upon *The Two Cultures and the Scientific Revolution*.[4] Snow lamented the fact that science had yet to become the foundation for our general culture and way of life He took it very much for granted that science deserved this position. And he did not doubt that science was capable of filling it; after all, science was already the foundation of a viable and vigorous culture and way of life, that of natural scientists themselves:

> the conservatives J. J. Thomson and Lindemann . . . the radicals Einstein or Blackett: . . . the Christian A. H. Compton . . . the materialist Bernal: the aristocrats de Broglie or Russell . . . the proletarian Faraday: . . . those born rich, like Thomas Merton or Victor Rothschild, . . . Rutherford, who was the son of an odd-job handyman. Without thinking about it, they respond alike. That is what a culture means.[5]

The trouble, as Snow saw it in 1959, was that although the scientists responded alike, the rest of us responded differently, in a way unsuited to life in a science based society. Instead of turning to scientists as the natural intellectual leaders of society we shunted them off from the mainstream of our social life into a kind of internal exile, an agreeable and privileged ivory tower existence, leaving the rest of us incompetent and vulnerable amidst the products of our increasingly technological age. For leadership we were distressingly prone to look to 'literary intellectuals', a 'coterie of natural Luddites' suspicious of, or even hostile to science, and possessed, by accident of history, of considerable political power. Between the literati and the scientists there loomed an immense cultural divide, reinforcing the incomprehension and distrust felt on both sides. The long-term consequences of this divide, Snow felt, could only be disastrous.

His cure lay in the use of education. The culture of science had to become the general culture; it had to suffuse through the whole of the political and administrative elite, and diffuse down into the fabric of everyday life There was a crying need, he insisted, in his inimitably parochial style, for

> as many alpha-plus scientists as the country can throw up. . . . a much larger stratum of alpha professionals . . . another stratum educated to about the level of Part I of the Natural Sciences or Mechanical Sciences Tripos . . . They will be required in thousands upon thousands . . . and last, politicians, administrators, an entire community, who know enough science to have a sense of what the scientists are talking about.[6]

Snow's views stirred up an intense and prolonged controversy. Not surprisingly. His was no mere plea for more scientific manpower. It was a basic statement about the kind of society we should seek to live in and the form of culture it should rest upon. It was, moreover, perfectly straightforward in what it implied for the distribution of social standing and political power. Scientists were not to be trained to be 'on tap', but educated to be 'on top'. They were to be active agents in political institutions, decision takers, wielders of authority in a wide range of contexts. And, needless to say, if scientists were to acquire power and authority on a great scale, as much of what Snow had to say implied, then others had, at least to some extent, to relinquish it.

To anticipate a later theme, Snow can be thought of as an advocate of a technocratic society, a society ordered around science and technology and dominated by scientific elites. This is why he was criticized so intensely and at the same time supported so vigorously. He had given clear expression to one possible vision of the role of science in society. Nobody disputed the immense role that science had come to play and had to continue to play in modern economies. But there was bound to be dispute as to whether science should remain firmly defined in the role of servant, or whether it should be allowed a much more pervasive and fundamental significance. And Snow, despite some hedging in his writings, was rightly perceived to represent the latter view.

MODERN SCIENCE

In many ways, modern societies have moved in just the direction that Snow advocated; indeed they had been systematically moving in that direction long before he wrote. The social standing of scientists and the cultural significance of scientific concepts and scientific knowledge have continued to increase. External opposition to science has continued to decline; amongst our political and intellectual elites it has all but petered out. As decade has succeeded decade everyday ways of thinking and accepted common sense knowledge have become recognizably more scientific: through the efforts of the schools and other educational institutions, and latterly through the mass media, they have been suffused with the idiom of science and to some extent with its knowledge and competences as well. A good proportion of what today is common sense knowledge is the scientific knowledge of yesterday in another guise.

Yet, despite all this, what Snow sought has not been realized, and will not be realized. Science will not become the basis of our common culture, pervading every aspect of our life and activity, providing common attitudes and sentiments, allowing us all to respond alike. Science education, although it will no doubt continue to bring benefits, will not bring quite the benefits for which Snow hoped.

Think of how things stand at present. It is true that scientists now have as much respect and influence as Snow could have wished, and that our culture is more strongly oriented to science than ever before. But for all that the perceived gap between science and everyday life has never been wider. Most people see science, quite rightly, as an activity beyond their understanding. And very many have in any case not the slightest interest in understanding it: many of the most popular newspapers and magazines devote more space to astrology and horoscopes than they do to natural science and its results. Although our everyday understanding has absorbed and assimilated much that was science in the past, it has become more and more distant from the science of the present. Science has moved on far too rapidly for our everyday understanding to keep up. And it has moved on in a way which actually makes it pointless for our everyday understanding to seek to keep up. A chasm between science and commonsense is a direct consequence of one of the fundamental tendencies in the development of modern industrial societies, and only the most revolutionary and far-reaching kinds of social change would begin to eliminate it.

I am referring to the tendency to specialization and an ever-increasing division of labour. We are all very well aware of this, and of the great increases in productive efficiency which it has made possible, as for example in the use of the assembly line system. But it is important to remember that the all too familiar division of physical or manual labour symbolized by the image of the assembly line has been paralleled by a division of intellectual labour, no doubt also inspired by the quest for greater efficiency. Certainly, the history of science is the history of a continual move to higher and higher levels of division of intellectual labour. Early in the nineteenth century science became clearly demarcated as a specific occupation. Research was established as the exclusive preserve of specially trained, formally qualified professionals, and a strong distinction was made between these professionals and outsiders. This was the first, and the most crucial, divide. But it was rapidly followed by a long series of further divisions within science itself, as the range of separate scientific disciplines emerged and established themselves in the course of the century. By the beginning of this century physicists, chemists, astronomers, geologists, mathematicians and many others could all be treated as members of distinct occupational groups, each with its special forms of training, formal qualifications, societies, journals, and all the various further embellishments of a learned profession. And since that time specialization and differentiation have continued apace, so that although the scientific discipline has remained the basic unit for most teaching purposes, the specialty within the discipline, or even the sub-specialty, is often the major point of reference for an individual scientist. Many scientific specialties now possess the kind of professional institutional apparatus that would once have done an entire discipline proud.

Specialization and differentiation on such a scale as this radically alter the activity and the experience of the individual scientist, and the way that he is perceived. Save for humour or irony we should scarcely ever refer to a typical modern natural scientist as a 'philosopher'. To reach the highest levels of his profession he must spend many years mastering a particular limited body of knowledge and competence, specifically related to the particular kinds of professional work he will later carry out, as a researcher and as an expert in his field. He needs know-how, techniques and methods, competences in particular kinds of manipulation, experimentation and calculation. And precisely because he will acquire this kind of knowledge and put it to good use, the rest of society has no need of it, and can afford to leave it out of account. Which is just as well, since what takes the scientific specialist many years and much difficulty to acquire could in no way be adequately assimilated into our everyday understanding. Modern scientific knowledge, the knowledge produced, preserved and employed by scientists themselves, has become packaged and organized to serve as the essential technical equipment for a set of particular professional roles, without regard for its relationship to our common culture and everyday understanding.

Needless to say, this development has implications for the way that scientific knowledge is transmitted. As science has become a range of specialized professional activities, so the higher levels of science teaching have become forms of specialized scientific training. It may perhaps be that in our schools, or in some of them, students are still educated in science in the full and proper sense of the term. But in the universities, and equally in other institutions of 'higher education', training is the name of the game.

Perhaps this statement is too extreme. There are, of course, many academic scientists who take a very broad view of their responsibility to their students and hence of their teaching obligations. And even now when scientists stand to gain a great deal in terms of political and financial support by obsessively stressing their economic relevance and their role in producing 'trained manpower', few of them would regard this as a sufficient account of their teaching activities. What a scientist would say that his concern is solely and simply to train his students, and that those of them wishing to become educated should become so on their own account by other means? Nevertheless the generality of modern science degree courses are not primarily intended as routes to enlightenment or expansion of the intellect. Their paramount concern, as indicated by what is central in their examinations and assessment procedures and by what consumes the major proportion of laboratory and lecture time, is to transmit to students a body of technical instrumentally-applicable knowledge much of which takes the form of specific skills and competences. And these skills and competences are selected by scientists, with remarkable unanimity, precisely as means to ends, and notably as a means to the end of successful technical performance in scientific research. A formal qualification in a natural science is now very often taken, much as a qualification in dentistry, as a sign that a certain set of skills have been acquired and a certain form of expertise is available.

Many authors, natural scientists among them, have compared the sciences to crafts which involve a repertoire of skills and techniques. And, extending the analogy, they have compared a training in a modern science with a craft apprenticeship.[7] This does indeed point up many of the key features of the experience. Like a craft apprentice a science student has to spend many long years in a subordinate role building up to a competent level of performance. This is not a time for the critical examination of scientific knowledge, or reflection upon its foundations: it is a time for assimilating what one is told and perfecting manipulative and calculative techniques. Teaching is likely to be to some extent dogmatic and authoritarian if only for the sake of efficiency, and teachers will generally expect what they have to say to be taken on trust. Above all, just as in a craft apprenticeship, an immense amount of sheer hard work is required, much of it the routine drudgery necessary to the acquisition and mastery of key skills and competences.

Think how, in most of the sciences, long hours of laboratory attendance are considered necessary, so that familiarity with instruments and their operation is acquired and the appropriate methods of analysing and interpreting their output are understood. Think too of the many hours often spent in 'examples classes', or in private study, working through set problems and examples one after another, endlessly, so that scientific concepts and symbols are understood as they must be understood *in use,* and mathematical procedures and techniques are acquired as they must be acquired, not abstractly as memorized series of symbols, but concretely as applicable competences. One writer has compared the exercises and standard examples worked through by science students with the finger exercises repeated over and over again by musicians, suggesting that in both cases the exercises are essential to learning, and in just the same way.[8] It is a fine analogy. It both highlights the effort and drudgery involved in the acquisition of skill, and the rewards to be expected later when effortless mastery is acquired and the opportunity for its creative deployment and use becomes available. Here at once is the sacrifice and the promise implicit in a modern scientific training.

The very high degree of specialization characteristic of modern natural science is often found deeply disturbing, not only by outsiders but often by natural scientists themselves. And it is easy to see why it should give rise to so much unease and anxiety. But the issues involved are very rarely faced squarely and unflinchingly. The achievement of modern natural science is founded upon its high level of specialization, and upon the readiness of individual scientists to concentrate all of their technical and intellectual resources upon a narrow front. One cannot take the achievement and leave the specialization, however fine the one and repulsive the other: they go together. One can reject both, of course, as some humanist critics of science with perfect consistency still do. Or one can accept both, although it is interesting to find that very few writers are prepared to defend specialization and an intense division of labour: even as science and society have continued to specialize at an extraordinary rate the consensus has been, at every point, that specialization has gone too far.

Specialization is accepted today as a necessary evil. As a means to an end it is marvellously effective, as the great economist Adam Smith brilliantly demonstrated over 200 years ago. But, as the basis for an individual life, specialized training seems to have no defenders. I am inclined to think that it should have, but I must admit that powerful arguments lie on the other side. It is very difficult, for example, to argue formally that something may be good and desirable when in its purest and most concentrated form it is an obvious evil. And an extreme manifestation of specialization is just this, an obvious evil. If a human being exclusively subordinates himself to just one end or one activity, he effectively sheds his humanity. This is intuitively clear to us. We actually use an image drawn from science to stereotype the evil in our informal thought – the stock image of the mad professor. And the realities of science do on very rare occasions approach this pathological extreme. A very few science degree courses, for example, expect a degree of obsessive involvement from students which amounts to mental hara-kiri.

Nevertheless I think that specialized training might be defended as part of the basis for a good life. It is indeed never perceived in this way in the context of science. But we do tend to perceive other highly specialized occupations rather differently. In our routine awareness of writers, painters and sculptors, composers, musicians and instrument makers, even, in a few cases, athletes and sportsmen, we seem perfectly willing to recognize that the possession of highly specialized skills and a deep devotion to just one narrow area of human endeavour can actually be a constitutive part of a rich, full and good life. We understand that people in these occupations can be complete human beings who draw deep satisfaction from their individual skills and at the same time make important contributions to the benefit of society as a whole through their use. Implicit in this informal understanding is an awareness that specialization may be a good. And however this awareness is rationalized, the rationalization must surely relate to the scientist as much as to the composer, the artist, or anyone else. The puzzle is, I suppose, why our routine stereotypes of specialized individuals vary so much from one occupation to another.

Very often, of course, anxiety about intense specialization is not related to its direct effects upon individuals but to its supposedly adverse consequences for society. The fear is that people generally may become too far distanced from the dominant forms of knowledge, those which inform key political decisions and legitimate them once taken. Widespread apathy and withdrawal, or even active disaffection and a threat to social stability may then arise. Needless to say, those who accept this view do not seek to do away with specialized scientific expertise altogether. Rather, they stress the need for that expertise to remain culturally connected with the rest of society. They sometimes argue, for example, that in a science-based society general education should be very heavily oriented to science, so that as much of the general population as possible is in some degree scientifically literate: this is Snow's dream of 'an entire community, who know enough science to have a sense of what the scientists are talking about'. But they also sometimes argue, conversely, that scientific specialists should be broadly educated as well as thoroughly trained, so that our communities of experts remain capable of easy communication with the rest of society and sensitive to its needs and attitudes. Implicit in both arguments is a conviction of the need for a high degree of continuity, coherence and consistency in our culture as a whole.

Arguments of this kind deserve to be taken very seriously indeed. Certainly, in the past it was very much taken for granted that a society had to base itself upon one single dominant set of symbols, images and beliefs, and that too much cultural diversity led to social disintegration. 'One ruler, one religion' was long the operative maxim in Christian Europe, and punishment for heresy was more severe than that for even the most heinous of secular crimes. It is, however, extremely difficult to decide what weight to give this form of argument in the present-day context. How far social stability depends upon cultural coherence, and whether the one could survive the demise of the other, is practically impossible to say, because there is a further factor making for social stability once a society has differentiated and specialized.

Specialization, whether by an individual or a group or an institution, is not merely the heightening of skill and competence in relation to a narrow range of tasks. It is also a loss of competence in relation to other tasks. It involves doing many things badly or not at all in order to do other things well. The outcome of this is that the whole society gains enormously in overall efficiency, but its individual members lose versatility and self-sufficiency. This indeed is our modern condition: we live in the most powerful societies in the world, but as individuals we are babes in arms compared to members of the most primitive societies, all of whom, unlike us, are generally capable of maintaining their own independent existence. We are utterly dependent on others in a way that they are not.

In a highly specialized society, intense relationships of interdependence come into existence between one part and another, and these relationships themselves serve as a kind of social cement. One group or faction or institution cannot turn too strongly upon another without at the same time hurting itself. Substantive agreement based upon shared understanding throughout the whole of a society is therefore that much less important, since disagreement is likely to lead not to fission or internecine strife but to compromise and some sort of *modus vivendi.* No doubt this is why modern societies can afford to be cheerfully sectarion in a way that earlier ones never felt able to be. Where there is interdependence there is that much less need for cultural uniformity.

Clearly, interdependence cannot altogether replace cultural uniformity as the basis for social stability. But interdependence does seem to be the more important factor as far as an understanding of the entrenched and secure position of modern science is concerned. Science has established itself and stabilized its position through its involvement in a vast network of relationships of interdependence much more than through any general diffusion of scientific ideas and attitudes. Our recent history is at one and the same time the history of the growing gap between science and the rest of society, the growing interdependence between science and society, and the growing dominance of science within society. Throughout this century science has established its position by the development and consolidation of strong relationships of interdependence with all our most important social institutions, in the field of technology, in the economy, in the military context, in government and the sphere of politics. If it is correct to speak of the existence of a military-industrial complex, as President Eisenhower apparently felt it was, then modern science is right in there amongst it.

Even the most esoteric fields of science and the most academic of scientific institutions are now firmly connected into this system. They act, in any number of ways, as developers, testers and suppliers of technical expertise. Perhaps the least important way is through the ready availability of their research literature, available to all, and thoroughly scrutunized by scientists and technologists in industry and elsewhere. More important is the availability of themselves, as consultants, advisers, entrepreneurs, bearing their know-how into a great range of contexts, putting it to use, and sometimes permanently implanting it. Possibly more important still is their output of trained students, moving into work: these are carriers of all kinds and levels of competences; basic highly generalizable skills applicable to a great range of technical and semi-scientific tasks; the specific competences of particular scientific disciplines with recognized practical value; new knowledge and new technique hot from the research laboratory and newly written into the textbooks – or perhaps only into lecture notes and laboratory handouts. All these are borne off, year after year, by generation after generation of students, into situations where their potential utility may be realized.

Nor is this invaluable output from academic science at all accidental or incidental. We tend to think of the academic world as the home of pure disinterested research, and so indeed it is. But it would be quite wrong to think that this is all it is. Indeed, it may even be wrong to think that this is what it mainly is. I have only to pick up the internal telephone directory of my own university to find that a good half of its scientific staff are explicitly listed under the rubric of an applied science (in which I include engineering and medicine), or as members of applied research units or similar establishments. And this is to make no mention of the many important applied projects which I know to be under way in the traditionally labelled departments, or of the interesting fact that the size of these departments and their output of students is frequently strongly related to the demand for those students in the economy. Nor is my own university particularly unusual. Developments within the ivory towers of the academic world have long been conditioned, not by the frenzied utilitarianism which seems to have afflicted many British scientists over the last five years or so due to particularly heavy pressure from government and politicians, but by a gentle yet prolonged interaction with the economy. Although academic science has been well provided with resources, much of which has been made available without visible strings attached, scientists have never failed to recognize that in the last analysis such a vast amount of support can only be sustained in return for services rendered.

There is considerable controversy about the precise extent to which science has become integrated into the economy and the overall system of production. Some people now see little point in distinguishing between natural scientists, and technologists and engineers, preferring to think instead of one large group of specialists generating and preserving our stock of potentially useful knowledge. But others point out that a tangible gap does still persist between these different groups, and that a good proportion of scientific research is still carried out without any regard for potential applications. This is true even of some of the most immense and expensive projects of 'big science'. Think, for example, of the thousands of millions consumed by modern fundamental particle physics, requiring as it does vast accelerators such as those at Geneva or Brookhaven. Particle physics is a field which offers little obvious prospect of economic pay-off, and which is certainly of no economic significance at present. (To be precise, it is of no economic significance on the supply side. It is, of course, a significant factor as far as demand is concerned: this is one of a number of features which have led some scientists to compare accelerator building programmes with the cathedral building programmes of earlier societies.)

But whether the scientific enterprize has completely sunk, as it were, into the sea of the economy, or whether it continues to float, like an iceberg, upon it, need not concern us here. The important point is that a great proportion, at least, of scientific activity is now coupled to economic and utilitarian objectives, so that science has become a component of inestimable importance in our system of production. In any number of locations, in industry, in government laboratories, in defence establishments, scientific research is routinely carried out to further quite specific practical ends. Most of the vast sums expended on scientific research can now be treated as investments. Such investments continue to be made because research has been seen to pay off in the past, and is accordingly expected to pay off in the future. What else would sustain a continuing world-wide investment in research and development (R & D) in excess of 100 billion pounds a year? Why else would developed countries continue, even through recent depressions, to invest between 2 and 3 per cent of their GNP in R & D?

Such sums continue to flow because R & D is thought to deliver the goods, to return a good yield on capital invested. And no doubt it does so. But it is very important to make a distinction between yields on investments and real human benefits, and not to allow one's thoughts to run too freely from the one to the other. In saying this I do not want to deny the immense real benefits which have derived from scientific research: in health, food production, energy use and many other contexts, they are only too obvious. But I do want to warn against thinking about real benefits in too short-sighted a way.

Consider the fact that of all the money spent on R & D a third at least is directed either to military projects or to projects solely designed to enhance national prestige – for example, the project which enabled a man to take a walk on the moon. Now science has certainly made a splendid contribution to fulfilling the military and the prestige objectives of particular nations: in this sense it has yielded the needed goods. But these specific needs which modern science has fulfilled only existed in the first place because modern science existed, and because it was in the possession of a number of competing countries. What has been done in recent years using space modules, Polaris missiles and the rest, was done in earlier times, every bit as well, using woad and spears. In international competition countries feel that they must use the best resources available; but whatever those resources happen to be they are always good enough. When spears and woad were the best, they were good enough. But now we have science, and therefore we needs must use science. A third of our R & D effort, therefore, is devoted to tasks which only the existence of science has made meaningful. It would appear that far from real human benefit springing from this sector of scientific activity, it merely represents science neutralizing the effects of its own existence and consuming immense resources in so doing.

I should add that those who like to defend the massive appropriation of scientific manpower for purposes of security and prestige often paint a rosier picture of the consequent state of affairs. True, they argue, all these scientists are trying to develop materials, processes, techniques, gadgets, for military applications or for use in space travel. But remember that this keeps them in employment. And remember too that all the things they find out and invent can be put to other uses, besides their primary military or space-related use. Defence programmes and space programmes sustain high levels of scientific and technological research from which we all indirectly benefit. Innovations spin off elsewhere. Research is done which benefits all of industry (and what a crafty way this is for governments to subsidize research, and thereby give their own industry a cost advantage over that of other countries).

The argument is absolutely correct. Research on one front does have useful applications elsewhere. Innovations do spin off. Scientific work frequently has far more significance and utility than its producer ever imagines. If we are prone to worry about the levels of military research expenditures or the amount of resources which goes into space research, we should remember the non-stick frying pan, the carbon-fibre racing bike and countless other similar spin-off innovations, and take comfort.

SCIENCE, TECHNOLOGY AND ARMAMENTS

I have not exaggerated the proportion of R & D work which is directed toward the requirements of security and prestige. Needless to say, to arrive at a precise estimate of the proportion one has to decide what to count as research and what as development, what to count as related to military concerns and what not, whether to count only direct relationships or to include indirect ones as well, and so on. And all such decisions can be called into question. But I think it is better to duck these technicalities and simply to take the proportion as a third, in the knowledge that most methods of calculating current expenditures and their distribution would yield a slightly higher figure.[9]

If discussion were to be limited to Britain alone the proportion would certainly rise significantly. Britain is one of a few countries which have chosen to invest particularly heavily in military research and development, and British government R & D support is particularly strongly biased in this direction: more public money is currently spent on military R & D than on all other R & D objectives combined.[10] So strong is this bias that it has given rise to a considerable degree of anxiety, based entirely upon hard-nosed economic considerations. Money which goes into defence-oriented research is money which fails to go into more straightforwardly useful industrial research. Similarly, outstanding skill and talent creamed off into the one is talent which cannot be put into the other. Thus, resources and manpower of rare quality become a burden for the rest of the economy instead of an asset to it. And, given that defence-oriented work is so often feather bedded, insulated from the competitive winds blowing elsewhere in the economy, and afforded the most lavish provision on the principle that no price is too high for security, the burden has a way of becoming onerous and ever increasing. This, at least, is an argument frequently put forward as relevant to understanding Britain's very poor relative economic performance, and the much greater growth rates in Japan and Germany where defence R & D spending is very much lower and industrialists have been able to make greater use of scientific talent to improve their international competitiveness.

I cannot judge this argument. Perhaps it is wrong. Perhaps a country can put an abnormally large number of its research eggs into the defence basket and still hope for a high degree of economic success. But this scenario is even more depressing than that of economic failure. In the long term research is a key to economic success, so success based on defence oriented research must in the main be success for the armaments industry. It must be success through increasing arms production and arms exports. But an economic success of this kind exacts a heavy political price. Constraints on foreign policy of a most unpleasant kind are liable to build up. A particular vested interest in arms production is liable to burgeon forth and exert pressure. And everyone is put into a position where they gain from increased international tensions and conflicts, and lose from their resolution. Surely it would be better to search for economic success via a move to a more balanced distribution of research effort.

Needless to say, however, the minor vicissitudes it implies for the British economy are not the major cause of concern about the deep involvement of science with the military. It is this involvement which has allowed the development and proliferation of nuclear weapons, and which remains as the backdrop to the present low-level confrontation between the major nuclear powers. It is this involvement therefore which has obliged us to live our lives in the presence of the continuing possibility of nuclear war.

For many people this is the overriding problem of our time, the problem which dwarfs all others, however extensive and involving they are. It is hard to disagree. Unfortunately, in its response to the problem our society is split right down the middle. For some of us nuclear weapons stand out, even amongst the rest of the grisly apparatus of war, as abominations. We should have no truck whatsoever with them. If we abandon our own nuclear weapons a nuclear exchange is so much less likely, and the threat thereby removed counts for more than any lesser evil engendered in the process: unlike other threats it is seen as a threat to the whole of life itself. But the dominant view is that the needs of the immediate situation demand nuclear weapons. It is claimed that the present distribution of missiles has created a balance of terror, so that restraint is pressed upon both sides by the knowledge that mutually assured destruction would be the outcome of any war. And it is claimed too, on the Western side, that nuclear armaments represent a comparatively cheap counter to the conventional might of the communist bloc, and that they thereby contribute to economic prosperity as well as to peace. These different styles of argument appear to carry weight with different groups of people, so that conflict persists and the opposing cases have become only too familiar as incompatible ways of understanding our present circumstances.

Circumstances, however, are liable to change, so that however urgent our immediate problems it is always wise to seek a longer view. There are indeed a number of long-term problems which can be expected to arise from the routinized military application of science. Earlier in the chapter I spoke of the institutionalization of science, and its subsequent establishment as a more or less permanent knowledge transformation system. This has obliged us to live on the basis of an ever-increasing, ever-expanding knowledge base, whether we would or no. We cannot realistically specify how much we would ideally wish to know at any point, taking account of our material needs and political sophistication at that time, and then push a button to switch the system off when we have reached the appropriate point. Nor can we easily forget or eradicate any given thing we have come to know, which we would rather we had never known. A society possesses its knowledge not primarily as a written record or a set of verbally formulated ideas; it possesses knowledge concretely in the form of knowledgeable people constituting knowledge preserving institutions. At the level of society in general, to lose knowledge requires a reversal of the process of institutionalization. The route back to ignorance, if anyone were to seek it, would lie not in the burning of libraries but in the execution of the scientific establishment, or in their acceptance of whatever might serve as the cultural equivalent of sterilization. In other words, the route back to ignorance, or even to the stable state in the realm of knowledge, is not open. The problem is how to live in security on the basis of proliferating knowledge.

Once routine connections are established between scientific research and the military, and extensive interdependence is created between the two, the problem becomes that of stable living with proliferating militarily applicable knowledge. And the possibility of backing away from this problem by dismantling routines and abolishing institutions is yet more utopian. Many of the vested interests linked with the fortunes of science now make common cause with the similar but even more powerful interests within and around military institutions. Together they constitute a formidable defensive front for the system as a whole. And within the system the demand for constant, urgent forward movement is enhanced and intensified by extremely powerful competitive pressures.

Above all, of course, there is the international competition which is the mainspring of the entire business. The 'logic' of such competition may take many forms: since 'they' have weapon X, 'we' must make a counter to X; since 'they' are in a position to make X, 'we' must be able to counter X; since it is theoretically possible to make X, 'we' must be able to counter it, in case it occurs to 'them' to make it; since God only knows what research might not make possible 'we' must do massive amounts of research just as 'they' are. All these and many other types of us/them arguments inform and legitimate decisions in the military sphere and keep production, development, and research at high levels. And of course such decisions are allowed for in the calculations of the other side, so that the antics of the two adversaries are mutually sustaining and mutually legitimating.

But besides the well-recognized role of international competition, competitive mechanisms lying wholly within the institutions of particular countries also play their part. It is natural here to think first of the market economy and the competition it engenders between particular firms. And indeed real competition for weapons' orders does play a part in maintaining high levels of R & D expenditure in defence oriented companies in Europe and the United States. But governments have a way of going easy upon their defence contractors and sparing them the worst effects of the so-called free market. In the defence sector, competitive tendering is the exception not the rule. Most work is done on the basis of cost-plus contracts which guarantee a profit for the manufacturer or developer, and actually encourage overruns in expenditure.

It is intriguing to note how governments currently stressing the relationship between economic efficiency and free competition nonetheless take care to attenuate or even eliminate that competition when the chips are really down, in the defence sector. War, or the threat of war, has a way of putting political convictions to the test. In both the world wars of this century governments of all persuasions responded to the pressures placed upon them by imposing central control and planning upon the system of production, and actually eliminating economic competition in the name of efficiency.

There is, however, another form of competition, just as extensive and just as intense, which keeps the ball rolling just as effectively. This is the political competition between the various bureaucracies, military and civilian, which are deeply involved in weapons procurement. Any large bureaucracy is a set of occupations, for the members of which it represents present rewards and future prospects. For long-serving members these represent the hoped-for return from the investment of a lifetime of effort. For senior members the entire bureaucratic organization stands as the embodiment of their power and influence, and its success or failure, expansion or decline, is also very much theirs. The bureaucracy is their empire, and their interest lies in expanding it. Empire building is a familiar aspiration which has been of major historical significance since the time of Ghengis Khan: only the means have changed.

Any particular nuclear policy decision at any given time affects not only the perceived military strength and flexibility of the country or alliance involved, but the size, influence and future prospects of all the various bureaucratic empires making up the relevant military-industrial complex. A new missile for the army is an increase in army jobs, army influence, army promotions and so forth. And conversely it is a relative decline in the parallel prospects for the navy and air force. Similarly it is a tempting morsel for established army suppliers, and for the areas and communities where the suppliers are established, and it is correspondingly less attractive for other suppliers and other communities.

If considerations of this kind are indeed important we should find them operative and evident in the texture of decision making. In particular, we should expect to find the testimony emerging from a given bureaucracy, whether concerning a specific weapon or an overall strategy, to be slanted in the direction of the vested interests established within it.

Concrete analysis of such testimony does indeed suggest that just this has happened. Certainly, a whole series of impressive studies by United States political scientists has revealed the great importance of 'bureaucratic politics' in the nuclear decision making of that country.[11] It is evident, for example, that the three armed services have each stressed the technical advantages of their own particular weapons systems, those under their control, and sought to increase their role at the expense of the others. And it is likely that the general strategic orientations favoured by the different armed services have also reflected their immediate interests. Thus, in the 1950s, when the United States navy knew that it could expect to receive submarine based missiles of low vulnerability but also of low acuracy, naval strategic thinking tended to stress the retaliatory deterrent use of nuclear power for which this type of weapon was well suited. In the air force, however, which was due to control more vulnerable but potentially more accurate land based weapons, weapons which offered the prospect of hitting not merely large cities but small enemy weapons bases, there was very much greater stress on the value of pre-emptive nuclear attack and a general strategy directed against military installations. Both services advocated the strategy which had the effect of increasing their own importance.

The ideal specification for a *deterrent* weapon has some strange properties. As well as being reliable and invulnerable, the weapon should be *inaccurate* and *slow* in reaching its target. If it is too accurate and too quick it may cease to deter, and actually incite an enemy to attack. Unfortunately, research cannot search for less accuracy and less speed. The thrust of institutionalized military research must be the other way, in the direction which diminishes any current deterrent effects in the nuclear arsenal associated with these technical parameters. Thus, it is not to be wondered at that strategies involving nuclear first strike have become more and more prominent in military thinking as the years have gone by, and that first strike capability is now considered a desirable feature even of submarine-based weapons systems.

'It may seem strange that nuclear war plans are fought over in much the same way as (say) departments in a local authority haggle over a budget, but it is so.'[12] This comment from a recent review succinctly presents the core characteristics of 'bureaucratic politics' and offers what is for most of us a vivid and immediate image of what it involves. All the various departments of a local authority share a common interest in a large budget: it is the cake from which they all take a slice. But when it comes to the actual slicing of the cake they are all thrown into competition with each other, each trying to maximize their own share. Similarly, many powerful organizations and bureaucracies share an interest in a large budget for 'the deterrent', but compete intensely with each other to increase their individual shares.

A very important part of this competition takes place at the technical level. The different interests constantly seek to improve the technical specifications of their own favoured systems, and to develop and modify them for new roles. This is a tremendous stimulus to research and development activity: new products, new prototypes, new possibilities, new basic theoretical and technical resources are all useful cards to play in the competitive game, where the prospect of a little more speed, or firepower, or accuracy, or elusiveness can make all the difference in the comparison of one system, or one development programme, with another. Indeed it has been suggested that concern with technical specification and technical sophistication has become a fetish in the Western military context, to the extent that it has become divorced from the realities of combat. Our military technology, it is said, has entered a 'baroque' stage.[13] It is a marvel of modern engineering in the tolerances required and achieved and the problems overcome in its production; and it is festooned with every kind of advanced technology. But although this makes it a striking symbolic representation and affirmation of certain important values and ideals, it does not necessarily make it good military hardware. Increased technical complexity can easily mean increased maintenance times, increased vulnerability, and greater unreliability in actual operational conditions. And it may make greatly increased demands upon the human operators of the technology, both in terms of their general intellectual powers and the amount of training they must undergo. There are now, apparently, some extremely 'advanced' Western weapons systems which are non-operational nearly all the time, and which nobody seems able to operate for the rest of the time.

Let me return to my principal theme. What I have tried to do in the last few paragraphs is convey a sense of the tremendous oomph behind militarily oriented research and development. This guarantees that the vast apparatus now more or less permanently devoted to such research is kept working at a high level of intensity, and that a rapid uptake of the research into military technology and military practice occurs. And this in turn guarantees a rapid rate of change in the overall military situation. But it is this situation which serves as the background for the political calculations which lead to peace and war, and as the situation changes so too must these calculations. Thus, in a nutshell, the establishment of defence oriented research makes it impossible to rely upon an enduring military balance. As things stand at present, means of maintaining peace will themselves have continually to develop and change: there is no one problem confronting us, but an endless series of such problems – endless, that is unless some very basic changes are made in the system of international relations and probably also in the institutional structures of the world's most powerful societies.

Imagine, strictly for the sake of argument, that the present military situation is indeed in some sense conducive to peace, and that the routine short-term calculations politicians make in that situation do not tend to take us to the very brink of confrontation. That is fine for the present. But the situation is changing all the time, because of competition, vested interest, and the ever-changing resources for competition and the realization of interests provided by research and innovation. And we have no way of knowing whether political calculation will continue to keep us far from the brink in the great variety of changed situations we shall face in the future. One society may see itself increasingly falling behind over the long term: should it launch a preemptive attack before the superiority of its enemy becomes crushing? A defensive umbrella may be produced, likely to be effective for a decade or so: is there an obligation to take advantage of this narrow band of time and precipitate at once what would only be even worse later? And so on. These, and no doubt far more testing questions, will face us, or rather our representatives, again and again, as time passes. And who can say how they will be answered?

Perhaps the most salient thing to bear in mind about our current nuclear arsenal is that it represents only a beginning. Whether one thinks of that arsenal as the product of science or as the product of industrialization it remains the outcome of processes which have been at work only for a couple of centuries or so. And the massive and systematic involvement of science in the military context has been with us for a mere half-century. Science has scarcely begun to show what it can do. Indeed, it has scarcely begun to show what it has already done. Much that is already well in train in research and development has yet to manifest itself. And even the half-forgotten research of past decades offers the possibility of frightening military applications, which have so far been left unexploited.

They say that a week is a long time in politics. Certainly, the kind of long-term problem I have pointed to here is not one which politicians know how to solve. Indeed, the long-term problem I have pointed to just is the politicians' solving, or attempting to solve, their short-term problems. There is some chance of course, that the problem will be solved by accident, that as we all make the end-of-nose calculations we have confidence in to solve our immediate difficulties, we will inadvertently, perhaps unbeknowingly, solve our longer-term difficulties as well. But no society has ever developed the means or the knowledge which would allow a systematic and reliable elimination of long-term difficulties of this kind, and it is hard to see how, in practice, the task might be undertaken. It makes sense, nonetheless, to attempt it, if we have a fancy for being anyone else's ancient history.

2

Science for its own Sake

RESEARCH

This chapter is about the community of professional scientists and its internal operation. It makes some mention of the historical development of that community, but in the main the focus is upon the present day. And here the discussion is really only relevant to academic science and disinterested, basic research, since it deals solely with the relationships between scientists themselves, and not those between scientists and other groups. This puts industrial and government research, which is now the major part of science quantitatively speaking, effectively beyond its scope, and it makes for an unduly narrow and idealistic treatment even of academic science, which is itself very intimately bound up with outside interests. Nonetheless, much can be learned from narrow and idealized discussions; they can serve to highlight important aspects of what is always an inordinately complex reality.

Typically, we are highly impressed with basic science and its research achievement. We think of it as a peculiarly effective and successful activity. What then is the key to the effectiveness, the secret of the success? Does it perhaps lie in the richness of individual talent that is channelled into science? Apparently not. What systematic studies there have been suggest that scientists are not, on the whole, individually remarkable: there is, apparently, little which sets them apart in terms of abilities, attitudes or personality traits. They are of course members of an elite occupation, and accordingly they possess the general characteristics typical of members of elites. In the United States and Europe, they tend to be white and male, to be above average in educational achievement, and to be slightly more fortunate than average in social background – all very much as one would expect. But in comparison with other elite groups there seem to be few marked differences in important characteristics such as intellectual capabilities, and indeed few marked differences of any kind.

Studies by psychologists do record large numbers of statistically significant but small differences between scientists and other groups, but it is difficult to know quite what to make of them. It is reported, for example, that natural scientists tend to be slightly more conservative politically and ideologically than are other elite groups, and slightly more mundane, orthodox and stable in their marital and sexual relationships. But this, if it is correct, must be set against the far more important fact that successful scientists have held all kinds of political opinions, have expressed practically every variety of philosophical and ideological doctrine, and have continued to practice practically every form of religious observance. And, similarly, successful scientists have manifested all kinds and degrees of sexual inclination and sexual activity: homosexuals, for example, have made magnificent contributions to science, just as they have made magnificent contributions to every other cultural tradition.

One particularly interesting and well substantiated finding about the individual characteristics of scientists is that, just like the rest of us, they do not perform well in tests requiring abstract, 'logical' reasoning.[1] Consider the following arguments:

(1) If the scientific hypothesis, H, is correct then the empirical event, E, will be observed.
(2) The event, E, is observed.
(3) Therefore hypothesis, H, is correct.

Clearly, this is an invalid form of argument: one cannot move from (1) and (2) to (3). It is, for example, obviously invalid to reason thus: if the moon is made of cheddar cheese then it will look yellow; the moon does look yellow; therefore the moon is made of cheddar cheese. Yet psychologists report that significant proportions of scientists accept this form of argument as valid; one recent study found a quarter of scientists making this mistake, another a third. Again, consider this argument:

(1) If the scientific hypothesis, H, is correct then the empirical event E will be observed.
(2) The event, E, is not observed.
(3) Therefore hypothesis, H, is not correct.

This is generally accepted to be a valid form of argument. It is actually the whole basis of the philosophical view that science proceeds by the falsification of theories and hypotheses. Yet one psychological study reports that almost half the scientists tested denied the reliability of this form of inference, and in another, slightly different, study, the vast majority of scientists denied it.

It is difficult to be sure how to interpret findings of this kind. It is always a mistake, when attempting an interpretation, to equate performance with capability, or to assume that behaviour in one context (testing) predicts behaviour in another context (science itself). But the results do help to disabuse us of the notion that science proceeds as the sum of the independent actions of individual scientists, and that the individual rational capability of scientists is sufficient to guarantee the results of science. Whatever plausible interpretation is put upon the above results, they do strongly suggest that isolated individuals would be far too prone to error for this to be the case.

The study of separate individual traits and capacities is not in fact the best way of attempting to understand scientists and what they do. No doubt there have been many important scientific advances made by extraordinary personalities with rare abilities. And no doubt to perform even the most mundane science some basic personality traits are necessary and some basic abilities, just as they are necessary for practically anything else. But it is in many ways more interesting to pass over these points, and to ask instead how generation upon generation of recruits to science, bringing with them for the most part nothing out of the ordinary in the way of individual abilities or attitudes, have without fail enjoyed resounding success, time after time, in their chosen occupation, and have added, in every generation, new achievements to what have gone before. Essentially, this is not a question about what individuals bring into science, but about what they find when they arrive. It is about the resources available to scientists and the guides and controls set upon what they do. It is, accordingly, a question about the social and cultural context within which scientists work, and about that work as collective, organized activity.

Now that science is an established occupation its successful accomplishments are well thought of in the same way as the successful accomplishments of other occupational groups. The sustained success of scientists in research, in the production of new knowledge, is in many ways analogous, when looked at on the large scale, to the success of car assembly workers in the production of motor cars. On the assembly line car production is a recognized objective; the relevant tools and materials are immediately available; and individual workers are bound together by communications, controls and incentives into an organized integrated system of production far more effective as a whole than the independently decided activities of the separate individuals involved would be. Similarly, in the science laboratory, research is a recognized objective; relevant tools and materials are immediately available; and individual scientists are bound together into an organized and effective system of knowledge production.

It will probably be the last part of this analogy which is the least easy to accept. If indeed scientists are bound together and organized, then clearly it is by practices and arrangements very different from those routinely employed in factories or on assembly lines. What then are these practices and arrangements? What does make science the collective activity which I have just claimed it to be? The answer must, I think, be built upon two central points. First, scientists in any field share a prolonged and intensive training which gives them a strong tendency to think and act alike. It is not just that they all receive the same body of accepted knowledge, and all acquire the same set of technical skills and competences. It is also a matter of acquiring common language and vocabulary, common conventions for the definition of units and scales of measurement, common signs and symbols, common taxonomic systems and schemes of nomenclature. Reality does not require these things: reality does not mind whether it is measured in feet or metres, nor does it insist that electrons are negatively charged and protons positively. But to be a scientist it is essential to assimilate and understand such conventions, and in scientific training they are rightly treated as of the utmost importance and drilled into students until their use becomes second nature. The result is that effective immediate communication is possible between the scientists in a given field, and that they all possess the capacity to learn from each other and the capacity to evaluate each other's work.

As well as these capacities themselves, however, there exist within the scientific community certain standardized procedures, certain routine ways of operating, which have the effect of ensuring that the capacities are put to good use, that individuals *do* rapidly learn from each other and *do* evaluate each other's work. The result is that the research of every single person is based upon the knowledge of all, and that the judgement of every single person is conditioned by the judgement of others. The effectiveness of research is thereby greatly increased, since the technical and intellectual resources available to each and every scientist are maximized, and the effects of his or her weaknesses or eccentricities are minimized.

What kinds of procedure do I have in mind, and how is it that they have the consequences which I have described?[2] Well, take for example the routine business of publishing a piece of research in a scientific journal. Let us imagine that a single scientist has carried out the research (start at A in figure 2.1) and is about to write it up and submit it (B in the figure). The scientist will be aware that a number of important conventions must be observed by a journal contribution. One such convention requires an impersonal idiom which omits any mention of the incidentals and accidents of the research process and concentrates entirely upon the results to be presented and their relationship to

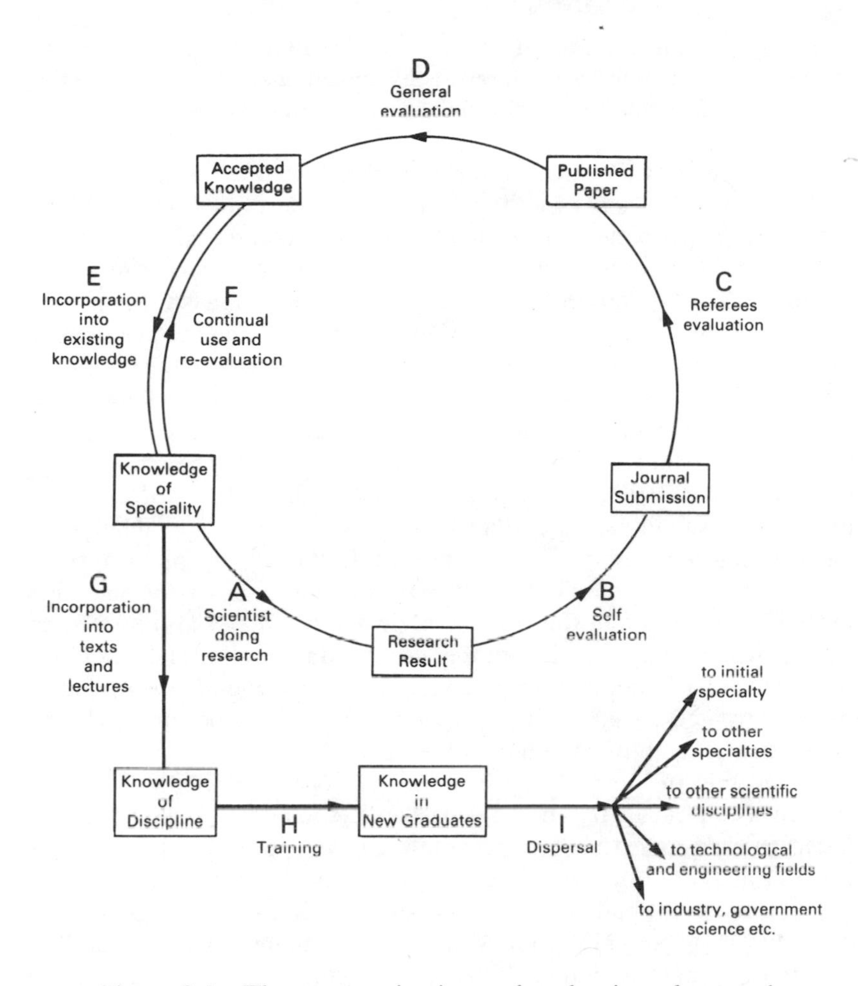

Figure 2.1 The communication and evaluation of research

particular scientific hypotheses. Another requires careful citation of the existing literature to indicate what is already known or accepted in the findings presented and techniques employed, and what is new and original. The scientist will also be aware that his paper will be scrutinized by referees, other scientists, his peers in whatever field of study is involved. And he will know of the likely views, convictions, standards of judgement held by those peers, and will recognize the need to take them into account. Accordingly, as he writes up his research the scientist will almost automatically transform it into a collectively acceptable form and re-evaluate and reconstruct it, perhaps even a number of times as one draft succeeds another, to take account of accepted standards. Already then, even before the research results have been made public, collective criteria of judgement and generally accepted standards of good practice are operative.

What of the further career of the research, once it has been submitted to a journal? One way of telling its story is to describe how it diffuses more and more widely through the scientific community, so that it becomes known to everyone else, a resource for everyone's research. Another way of describing the same process of diffusion is to speak of a long series of further assessments, which means that the research is effectively scrutinized by a significant cross-section of qualified opinion before it becomes established as generally accepted knowledge. The first stage of assessment/diffusion involves referees (C), specially selected for their detailed knowledge of the topic involved. They may accept or reject the contribution, or refer it back to the author for improvement and resubmission – so that it is reassessed and re-evaluated yet again by author and referees before publication. And publication is itself merely the prelude for further evaluation, as other scientists examine it for the first time (D). Only when the material has diffused widely, when it has been favourably cited and reviewed, and when it is clear that it has failed to attract adverse comment and criticism, does it become treated as accepted knowledge, to be incorporated into the main body of established findings of the field (E).

Nor is this the end of the evaluation process. The knowledge of science is empirically based and hence subject to uncertainty and always liable to revision. Even as accepted knowledge is used in further research it is continually reassessed in use and its status reconsidered (F). Science is not built like brickbuilders build a house, with each brick checked for shape and soundness and then permanently cemented into the structure of the building. Evaluation occurs again and again: every part of the structure of science is subject to continuing reappraisal, although of course some parts of the structure are scrutinized much less frequently then others.

It is now possible to see why the high tendency to errors of inference, and the generally low level of competence in formal logical argument amongst natural scientists is of no very great importance. Practically every single move in the game of research is played over again and again, in the first instance by the particular researcher, and subsequently by his peers. Because of this, single, one-off, individual judgements have little long-term effect. What matters is the tendency for a series of judgements to stabilize upon a specific outcome. This outcome will be the one devoid of error and idiosyncracy, the one which reflects the collectively accepted standards of the scientific community and a generally agreed view of how they apply to the case in question. Error prone individuals systematically organized and coupled together in their professional activities make a highly reliable, far less error prone, knowledge producing machine.

The routinized system of journal publication serves as an invaluable means of communication for scientists, and as a means of error elimination and quality control. And it itself is only a part of a larger system. It is easy to see that any routine practices which encourage the professional interaction of scientists, conferences and meetings, exchanges and visits, and so forth, will act in the same way. Scientists, we can say, have *institutions* which facilitate communication and quality control. A brief discussion of the journal system has indicated a little how such institutions operate and what effects they have. But there is the further question of *why* institutions such as the journal system operate, why scientists choose to behave in a way which sustains the system and keeps it in existence. And this is itself just a part of the very important and completely general question of what motivates scientists, what induces them to act in the way that they do.

Basically, this is a question about reward and incentive, and before attempting to answer it in the case of scientists it will be useful to consider a more familiar and mundane form of organized activity, and to ask how it is rewarded, what incentives are provided to keep it in existence. Let us go to the car assembly plant again. To assemble a motor car a vast range of tasks must be performed in a co-ordinated way by a large number of separate individuals. On the whole, these individuals do not do these tasks because they find them intrinsically satisfying. Very, very few workers gain deep satisfaction from screwing on the left front wheel, or fitting the right-hand indicator light cover. Individuals perform these tasks as a way of fulfilling various outside wants and needs: the possibility of fulfilling these wants and needs is the major incentive to work. But different individuals have different wants and needs, and hence find different things rewarding. The actual incentives to work vary from one person to another: one person may be moved by an overriding need to feed and clothe a family, another by a love of travel which he could never satisfy if unemployed, another by material possessions. Clearly, it would be wellnigh impossible to negotiate with a large work-force so that individuals worked as required in exchange for their particular needs being met, a trip abroad for the indicator light, a video for the left front wheel, and so on. There is a major technical difficulty in coupling particular tasks to particular wants so that everyone has a potent incentive to work.

Fortunately, however, there exists in all modern societies a marvellously useful invention which enables this problem to be dealt with. It is money. The use of money is of course an exceedingly complex phenomenon of truly profound interest, but basically it is operated as a general medium of exchange, as a currency. The possession of money is the potential possession of a great range of particular rewards – whatever rewards are priced in or obtainable through money. Money is a *route* to a great number of particular rewards. To offer a worker money is, as it were, to offer him his choice of particular rewards. Thus, the use of money in the car assembly plant couples the motivating power of a vast range of particular rewards to the task of encouraging required work performances. A single wage agreement replaces a great number of transactions exchanging particular performances for particular rewards. A magnificent simplication of exchange relationships is achieved.

Needless to say, once a monetary system is established people will see money in itself as a desirable thing: they will state that money is what they work for, money is what they want. And in a sense this is perfectly true. But it is true in a very limited sense. Very few people enjoy counting the stuff, or keep it under the bed in a sock as a prized possession. Some people are like this, of course, and the system would still work even if many people were. But on the whole money is wanted as a means to an end, not as an end in itself. Its attractiveness is a reflection of the attractiveness of the particular rewards which, as it were, lie beyond it. When people work for money it is the rewards lying beyond that are the ultimate incentives. But money is now a route to very many particular rewards, practically every possible particular reward, so that most people will work 'for the money' whatever it is that they might want. At this point, the particular wants that individual people have cease to be interesting in understanding why they work: all that matters is that they have wants of some kind. Whatever their individual wants, whatever their individual psychology and motivation, it will probably make them susceptible to monetary inducements. All we need say at this point is that work is related to a *reward system* using money as a currency.

When we turn from situations such as the car assembly line to what appertains in the academic research laboratory a striking contrast is immediately evident. It is not just that there seems to be less direct connection between tasks and monetary rewards, it is very much as if research, particularly judgement and evaluation, is kept as independent of financial incentive as is possible. Research is never anywhere done as piece-work, and in the realm of academic science there is no explicit connection at all between rate of pay and research produced. Nor is any payment made for the writing-up of research and its being handed over to a journal. Journals accept papers as gifts, just as they also accept the services of referees. Only review articles and popularizations are explicitly commissioned and paid for. Moreover, scientists themselves, even if not so much today as formerly, tend to be averse to direct connections between research and personal financial reward. On the whole, academic research, unlike the assembly of motor cars, keeps going without the direct involvement of money as an inducement.

This is not because scientists are altruistic and idealistic individuals, although many of them may be. It is because a different *system* of reward operates in science. Academic scientists are like many other professional groups in that they appear to be more immediately concerned with honorific reward than with financial reward. They tend to be particularly concerned to obtain recognition from their fellow scientists, and attach considerable importance to signs of that recognition, such as approving citations, formal honours, awards and prizes. The publication of one's research results is an accepted way of obtaining recognition; recognition can be expected to accrue to any published work which is at once original and important in the eyes of others. Thus, recognition serves as an incentive and reward for the execution and publication of original and significant research. It could be said that the scientist surrenders his original knowledge to the whole community in exchange for recognition by that community. The benefits of the transaction are obvious as far as the scientific community is concerned, but what do they consist in for the individual scientist?

It might be suggested that recognition – whether fame or simple respect, acclamation or simply the good opinion of others – is something much desired in itself by many or perhaps most people, and that the hope of it is what motivates many scientists. Perhaps this is true: I do not know. But to think of recognition solely in this way will not do. Recognition is not merely one possible form of reward for the scientist: it is implicated in the system of reward in a much deeper, more fundamental way. Even a scientist who was indifferent to recognition, even one who loathed and detested it, would be obliged to acquire it if he wished to have a successful scientific career.

The routine procedures of the scientific community are laid down in such a way that recognition is the route to all things. Does the scientist seek a research grant? His chances will probably depend on how well recognized he is by his fellows. Does he seek equipment, or more space, or the time and assistance of his colleagues, or simply to keep his colleagues off his back and lead a quiet life? Again recognition is the key. Does he seek personal gain, by a move to a better-paid post, or the accumulation of consultancies, or even by a skilfully timed leap out of research into science policy or administration? Even here the extent to which he is recognized by his scientific peers may be crucial. Like it or lump it, recognition is simply necessary. Whatever it is that the scientist wants, the way of getting it is via recognition.

Note that this discussion of recognition as a route to reward runs

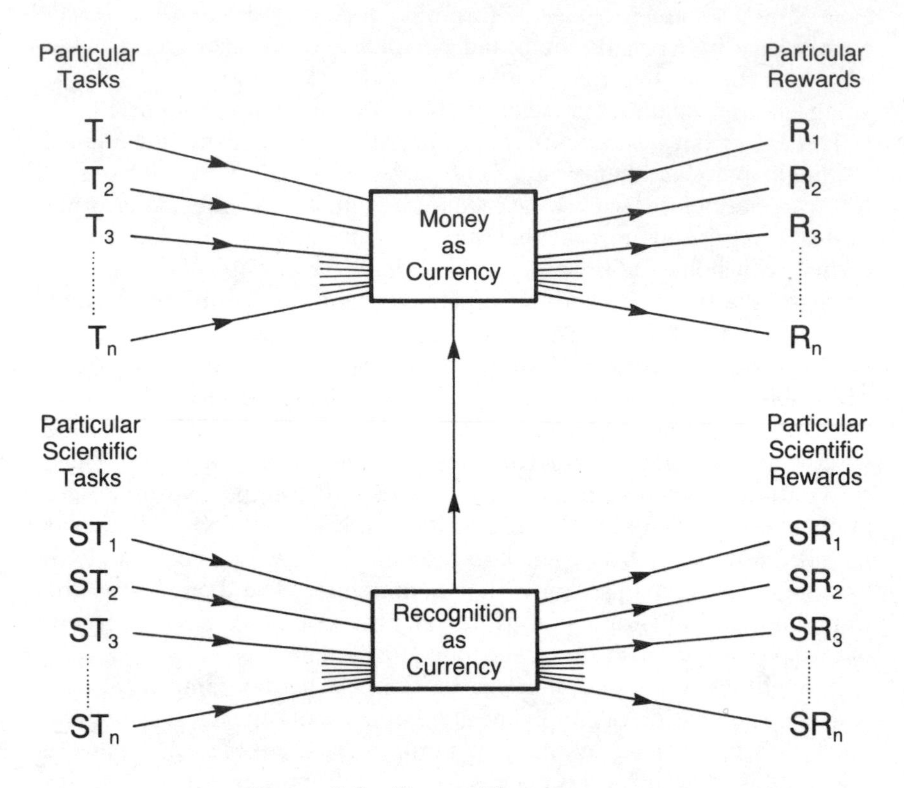

Figure 2.2 Recognition as currency

exactly parallel to the earlier discussion of money as a route to reward. And indeed recognition plays a role in science that is just like that of money in the car assembly plant. Just as money makes a great range of potential particular rewards into real inducements to car production, so does recognition make a range of rewards into inducements to research. It is now possible to give an answer to the question as to what motivates scientists and induces them to carry out and publish their research. They are motivated by a desire for recognition, which is only obtainable by doing and publishing research. But, it must immediately be added, they are so motivated only in the very narrow sense that car assembly workers are motivated by money. Recognition is sought after as the route to a whole range of wants. It is sought after, whatever scientists want, as the route to that want. Recognition is quite literally the *currency* of the reward system of the academic scientific community.

The use of a currency different from that employed in the main body of society allows scientists to maintain a considerable degree of autonomy, and particularly autonomy of technical professional judgement. This is probably the main reason why scientists, like a number of other professional groups, continue to accept and defend the systems of honorific reward which they have inherited and to avoid linking their most crucial professional activities to financial incentives. Let us examine how the use of a separate currency helps in the maintenance of professional autonomy.

In figure 2.2 the tasks $T_1 \rightarrow T_n$ at the top left are all the usual kinds of task which are performed as the general economy operates; among them are the various kinds of task involved in car assembly as discussed earlier. By performing any of these tasks an individual may gain money, and then exchange the money for any of the particular rewards the money can buy, $R_1 \rightarrow R_n$ at the top right. Money provides a route from any T to any R. In contrast the performance of any specifically scientific task, $ST_1 \rightarrow ST_n$ at the bottom left, yields in the first instance not money but the recognition of fellow scientists. This recognition can be used in obtaining specifically scientific rewards such as research grants and resources, $SR_1 \rightarrow SR_n$ at the bottom right, or it can be used as a route to the acquisition of money (as when one's publications qualify one for a better-paid post), and hence as a way of obtaining all the common rewards, R, which money can buy. For a scientist, therefore, recognition is the main road to all the rewards which science in particular provides and all the rewards which society generally provides, all the SRs and all the Rs. The need for recognition is therefore extremely strong, and since the granting of recognition is a matter for the scientific community it gives that community an extremely potent means of controlling individual members. The allocation of recognition is indeed the whole basis of the power of the community.

Finally, and most importantly, although recognition is a route to money, money is emphatically not a route to recognition. Whereas there are routine methods of exploiting one's scientific eminence to make money, there are no routine ways of exploiting wealth to obtain scientific eminence – whether for oneself or for others. Recognition is legitimately accorded only to qualified scientists and only for research. Thus there is no way for outsiders to deal in the currency of science and thereby to attempt to influence scientific judgements and evaluation. The recognition system remains wholly under the control of the scientific community: a strong taboo is observed against any action which might weaken that control. This is how professional scientific autonomy is sustained. Were the secondary currency, recognition, to be withdrawn and scientific tasks rewarded directly by monetary means, science would pass altogether out of the control of the knowledgeable into the control of the wealthy. So long as recognition remains the necessary first step to reward for a scientist, and so long as recognition has no price, a degree of scientific autonomy is maintained.

Scientists, it must be remembered, recognize a duty to make their knowledge available to their entire community, and there is generally a significant demand for that knowledge. Academic scientists meet the demand by acting as consultants, bearing knowledge to its users directly; and by teaching, so that knowledge is borne to users indirectly by newly graduated scientists (H and I in figure 2.1). Naturally, the users of knowledge try to influence the producers – to research in one area rather than another, to seek out economically significant topics, even to produce politically expedient results and avoid or disguise politically explosive ones. And, of course, this pressure has an enormous impact in the long term upon the nature and distribution of research, which is only to be expected given the immense financial resources of government and industry, the users of scientific knowledge, and the immense importance of scientific knowledge to these users. But if the power of money could reach directly into the core of scientific research and systematically condition scientific judgement and evaluation, the scientific enterprise would be thoroughly corrupted and its standing and authority would decline. Research to show that tobacco smoke does not after all cause cancer would be worth millions upon millions to interested parties in modern society. Results demonstrating that moderate doses of radioactivity are good for you would be almost priceless. The possible consequences of coupling research judgement directly to financial incentives and rewards are mind-boggling. And it is these consequences which are prevented, or perhaps we should simply say lessened, by the use of a secondary currency which is controlled entirely from within the scientific community and which can be earned only by members of that community.

The existence of the independent reward system gives the scientific community the opportunity to control its own core research activities, to maintain its own distinctive standards and paradigms, and to build up its own distinctive body of knowledge and competence with reference to those standards and paradigms. It allows scientists to accumulate findings and develop theories without undue regard for whatever economic and political concerns currently predominate in the wider society. Needless to say, it would be an ideal world wherein scientists made the most of this opportunity, and the history of science records example after example where they did not. But it is generally accepted that for the most part the factual claims made by scientists, and their ways of describing and interpreting events, are in some sense more detached and disinterested than are, for example, those of politicians and industrialists. And this relative disinterest, which science, and particularly academic science, is generally perceived to have, has added greatly to its credibility and the extent to which its members are trusted and respected in the wider society.

WHO COUNTS?

The existence of the recognition based reward system enables scientists to separate themselves off from unqualified outsiders, and to separate off their own relied-upon knowledge and technique from the possibly less reliable knowledge and technique of the outside world. It is sometimes thought that the system serves to keep bias out of science, and to allow truth to accumulate within its boundaries as a result of disinterested scientific investigation. This view, however, wherein the institutional arrangements of the scientific community serve merely as a protective shield around a rational, truth-producing activity, is far too idealistic. And it begs far too many questions – questions still unsolved about the character of rational activity, the nature of truth, and how far science actually does embody the one and produce the other. It is much safer to stay close to what can be observed, and to say simply that the system makes a strong distinction between insiders and outsiders, and between inside knowledge and outside knowledge. Insiders, scientists, have recognition and standing in science in a way that outsiders cannot hope for; inside knowledge is trusted and accepted in a way that outside knowledge (even 'correct' outside knowledge) is not.

Let us take first the distinction between insiders and outsiders. In science today insiders count and outsiders do not. What a qualified scientist asserts is almost invariably taken seriously, and often trusted to be correct just because of the reputation of the scientist who has asserted it; what an outsider asserts is generally not taken seriously, and is only actually accepted after it has been independently checked out by an insider. This is an important point. Think what it implies. For something to count as a scientific fact today it must be observed by a qualified scientist. Hence, the evidence for and against a scientific law or conjecture is the sum of what qualified scientists have reported one way or the other. And hence, the conceptions of the world currently established in science are largely built out of, and on the basis of, what qualified scientists have reported. We have here an account of science slightly but significantly different from the usual one. Scientific knowledge, we are now obliged to say, is based not upon experience as such, but upon the experience of the scientific profession.

In some sense, this statement must surely be correct, but it is very counter-intuitive. It is an established ideal in science that openness to experience should be maintained: a scientist is supposed to check knowledge claims directly, not by asking, 'Who says so?' But in fact, 'Who says so?' is a very important question indeed in science, one which scientists routinely ask, and need to ask, all the time. This kind of question does not exist in the traditional ideal of science because that ideal is not a realistic description of science as it is actually practised but more a statement of the standards which have priority in the last analysis. The ultimate court of appeal in science is direct observation. But the routine practice of science cannot involve that ultimate court all the time, just as the practice of law cannot involve the House of Lords all the time. In the actual everyday practice of science, 'Who says so?' is a question with a central role.

Naturally, in asking 'Who says so?', and then largely discounting what outsiders say, scientists hope to eliminate incompetent and spurious observations as well as biased and mischievous reports. It is interesting to reflect, however, that by discounting outsiders scientists also massively reduce the amount of information they have to deal with. Imagine an ideal society, without incompetence or bias or mischief, a society of reliable observers. Could scientists hope to deal with all the observations those observers might see fit to report? Could science continue without being highly selective even in the reliable observation reports to which it attended? More likely, by attending to them all it would go into overload and collapse under the strain. Imagine an individual doomed to remember everything seen and heard, every detail of every object, every aspect of every event. The way to madness would be wide open, and quickly traversed. One of our major assets as intelligent knowledgeable individuals is our ability to ignore and to forget. To lose this ability would be a disaster. We should be left with no hope of interpreting the world as an orderly, intelligible coherent reality. On a different level, the same thing can be said of science. Science must have a low sensitivity to the environment and a poor memory: most of the information impinging upon it, even good, reliable, information, must either fail to register or else be rapidly forgotten. Scientists must operate an information filtering system with a large rejection rate.

Naturally, scientists will prefer a system which retains as much of the really important information bearing in on them as possible. They will seek to filter information according to its reliability, precision, relevance and utility, so that the small amount of material they eventually concentrate upon is as rich in significance as possible. But science is finite in its resources. Just as it cannot deal with unlimited amounts of information, so similarly it cannot put unlimited amounts of effort into filtering that information. Science, therefore, must operate some crude and simple filtering systems, which work less than perfectly, but fast. It must take the chance that a certain amount of valuable knowledge may be lost, and a certain amount of rubbish let through, in exchange for a filter system which rapidly cuts down the input of information to manageable proportions. To use a convention whereby only the reports of qualified scientists count is to use precisely such a filter system.

Concern with, 'Who says so?' is no modern development. Back at the time of the Scientific Revolution, well before the term 'scientist' even existed the question had a role. A fascinating example is afforded by the research practices of the members of the Royal Society in the period immediately after the restoration of Charles II in 1660. In this period, of course, science lacked the standing and respect which it enjoys today, and its practitioners had to fight hard to establish their credibility. In these circumstances, great play was made with the central ideals of science, and with the way that its knowledge was based upon direct experience and not upon hearsay. Members of the Royal Society appealed to the facts of nature as phenomena which all men alike might perceive and about which all men alike could be certain. They reckoned that if science could establish itself purely on the basis of fact it could achieve the credibility it was in need of. Hence when they performed experiments they went to great lengths to isolate and identify the visible facts and to separate them (as they thought) from theory and hypothesis. Experiments were performed in front of witnesses and the testimony of witnesses recorded. Agreement amongst many witnesses was taken as a sign that what was agreed to was indeed a genuine fact, and not the spurious product of individual bias or individual incompetence. That every man in a group of witnesses would testify to having observed a phenomenon was taken to show that any man could observe the phenomenon, that the phenomenon was really there, a fact of nature.

Yet even in this context, where the direct observation of nature was given such immense explicit stress, the matter of 'Who says so?' remained important. Witnesses at important experiments were not randomly selected members of the population. They were invariably persons of high social rank, gentlemen or better; and preferably gentlemen with some relevant competence. 'Virtuosi' was a term much used at the time to describe such people: Robert Boyle wrote of performing his experiments 'in the presence of an illustrious assembly of Virtuosi'.[3] Indeed when important experiments were undertaken Boyle would not only assemble witnesses, but record their names and their qualifications as material relevant to the credibility of their testimony and hence to the standing of the reported experimental results. Witnesses had to be credible witnesses, for which it was essential that they possessed requisite social and scientific standing. Just as witnesses in a court of law were more or less credible according to who they were, so, it was recognized, were witnesses to a scientific experiment. Anybody at all just would not do as a witness.

At this time, of course, a person's trustworthiness was formally associated with social rank. The higher one stood in society the more educated and perceptive one was thought likely to be, and the less credulous. And the higher one stood in society the more a lapse into bad faith was liable to stain one's honour. On both counts, the higher one stood the more reliable one was thought to be as a witness. Whereas 'the common people' were in no way to be trusted, 'the word of a gentleman' had to be treated seriously. Indeed, it is tempting to suggest that the whole system of quality control in science may have developed from what initially was a very simple system heavily dependant on 'the word of a gentleman'. Our modern understanding of what is required of a qualified observer may be thought of as derived from this earlier notion, in a development where more importance has gradually been given to training and professional status and less to general social standing. Science, we might say, has always had to place people upon a hierarchy of credibility in order to filter knowledge-claims and ignore the less plausible ones. But whereas initially the generally accepted social hierarchy was used as the hierarchy of credibility, today there is a specialized hierarchy defined by professional qualification.

Throughout the transition one basic theme would appear to have remained as a constant. 'The common people' as was said then, 'the general public' as we might say now, were not to be trusted. They were the bottom of the hierarchy of credibility whether it was defined in general or in specialized terms. Their observations and their claims to knowledge were the least reliable, the most open to suspicion. An amusing illustration of this is provided by a scientific controversy which occurred in the eighteenth century, in a period when the gap between expert practitioners of science and lay persons had increased, but when science was still far from being clearly demarcated as a profession. The controversy concerned the existence or non-existence of meteorites, solid bodies of extra-terrestial origin.[4]

The meteorite controversy involved parties in many European countries, but much of it centred upon the French Academy of Sciences, whose members we might today reasonably refer to as scientists but who referred to themselves as 'savants' and 'philosophers'. Throughout most of the eighteenth century the savants refused to accept the existence of meteorites, and rejected numerous eye-witness reports of their fall. Only at the very end of the century did the members of the Academy change their collective opinion and give credence to the acceptance of meteorites. Since today we routinely accept that such objects exist it is natural to ask why recognition of the fact took so long.

It is important to recognize first of all that the connection between meteors and meteorites was not recognized at the time. When an extra-terrestial body enters the earth's atmosphere it becomes a fireball visible over a very large area. Many savants had themselves witnessed such events, meteors. But it was not clear that the events were caused by solid bodies making incursions into the atmosphere. Only if the remains of a solid body could be found, a meteorite, at the end of a meteor's path, would there be evidence of this. And whereas many people saw meteors, very few found meteorites. To see a meteor one had to be within, say, tens of miles of it. To find a meteorite one had to get within yards of it. The difference is enormous. And, in any case, we believe today that most meteors burn themselves out before completing their passage through the earth's atmosphere and that only a very few leave meteorites as tangible indications of their existence.

Naturally, meteorite finds were made in country areas, not in the very tiny percentage of land space taken up by the cities in which the savants typically lived. Observations of falls and associated finds of meteorites were made by 'the common people', and it was precisely this which guaranteed that they would largely be discounted, even though over the years a considerable number of such reports accumulated. Many testimonial accounts of meteorite findings as well as a number of putative meteorites were put into the hands of members of the French Academy through the eighteenth century, and there had been similar submissions to learned bodies in other countries. But the first thorough scientific analysis of such a find was only made by the Academy in 1772; and even then the committee charged with the investigation, headed by the eminent chemist Lavoisier, reported negatively, stating that the submitted objects were terrestial in origin. It was to be over 20 years before this conclusion was seriously thrown into doubt, and more than 30 years before the Academy finally accepted that meteorites were genuinely extra-terrestial objects.

The continuing reports from 'the common people' simply were not taken seriously. Even the existence of a considerable and ever-increasing number of such reports did not carry weight amongst most of the savants. As one of them put it:[5] 'everybody knows that thousands of absurdities have been certified by thousands of witnesses of this nature.' Amongst the savants the credulity of the multitude was legendary. When in 1790 a large fall of meteorites was witnessed by about 300 people in a small community in the south of France, the first response of a professor acquainted with the news was evidently no more than amusement. The professor was surprised when later he received a legal deposition from witnesses, including the mayor and the town attorney, together with samples of the meteorites themselves; but for all that he remained completely unconvinced by the report. And when he passed on the report to be noticed in a scientific journal it stimulated a bitter editorial commentary.[6] 'If the readers have already had occasion to deplore the error of some individuals, how much more will they be appalled today seeing a whole municipality attest to, consecrate, by a legal protocol in good form, these same popular sensations, which can only excite the pity, not only of physicists, but of all reasonable people.' Clearly, if every savant had taken this kind of attitude science would have been able to learn of nothing at all from 'the common people'. Every last trifle of information which they sought to provide would have been filtered out.

In fact, of course, the savants did revise their views about meteorites; otherwise I could not have described events in the way that I have. And the change of mind of the savants was actually in part a response to the reports of unqualified witnesses. The work which initiated the change of mind was a book by Chladni, a respected savant, published in 1794, which assembled and compared a number of reports, noted the similarity between the various finds involved, and postulated a connection between the attested phenomenon of meteors and the objects subsequently to be known as meteorites. Chladni's book was written entirely on the basis of library work: the various reports stored in public libraries, all originating initially in 'the common people', proved a sufficient basis for Chladni to assemble a powerful if not conclusive argument.

Chladni was not just receptive to, but actively searched out, observations witnessed by 'the common people'. But he was well aware that he was risking his intellectual reputation in doing so: he actually refrained from quoting other scholars who accepted his claims, for fear of thereby bringing them into disrepute.[7] And even though Chladni made a strong impact, it was not his book which, for most savants, conclusively settled the controversy. The crucial evidence here appears to have been provided entirely from 'reputable' sources. In the late 1790s detailed analysis of many of the diverse stones claimed to have fallen from the sky revealed a remarkable uniformity in their chemical composition: they frequently proved to be iron, with a high nickel content never so far found in natural earthbound nodes of iron. It seemed most unlikely that a number of independent spurious finds made by 'the common people' would all alike manifest this most unusual and difficult to identify characteristic. Here the inherently unreliable testimony of the credulous multitude was supported by evidence originating from within science itself. Similarly, when in 1803 a massive fall of stones was reported within 70 miles of Paris, Jean-Baptiste Biot, a highly respected savant, was able to travel to the scene and undertake a direct investigation. It was his testimony which effectively silenced the last remaining sceptics and brought the controversy to an end. 'The common people' had usefully informed and influenced the savants, but the 'scientific observations' which settled the controversy were the observations of the savants themselves.

In this case, the quality control system of science was over a long period of years largely impervious to observations originating in 'the common people'. It was extremely difficult for such observations to pass through the established filtering system: they were almost certain to be ignored or forgotten or discarded. As a result the acceptance of an important body of knowledge was long delayed, and erroneous views about the nature of meteors and the possibility of extra-terrestial intrusions into the ambit of the earth were allowed to persist uncorrected in the context of science itself. This makes it very tempting to say that the scientific filtration system was far too demanding, that the savants should have remained more open and receptive to the reports and observations of outsiders. Indeed the whole story could be used as a moral fable to support certain traditional views of how science should operate: see what happens, it might be said, when scientists become dogmatic and close their minds, when they cease to attempt to judge observations on their intrinsic merits.

Needless to say, the situation is actually far more complex and difficult than this approach recognizes. No doubt the filter system did reject valuable material in this case. But such a system cannot simply be judged after the event on the basis of one instance of use. One has to think of the general implications of its design, of what would have happened overall if the filter had been made less retentive, less intolerant of reports from 'the common people'. Savants might indeed have done well to heed the widely accepted popular belief that solid objects fell from the sky. But should they similarly have taken seriously the view, also widely accepted, that such objects had supernatural properties, or the view that they could be used to alleviate the traumas of childbirth? And how seriously should popular testimony have been taken when supposed meteorites turned out, as some did, to be fossils?[8] It seems reasonable to suggest that if all that had previously been put down to the 'credulity of the multitude' had instead been taken seriously, a massive flood of spurious testimony would have created enormous logistical problems for the receiving scientific institutions. It is very far from clear that the gain of genuine and significant information would have been sufficient compensation for the loss of expert time and effort caused by an inundation of rubbish. It cannot be said that the savants were 'right' in their general attitude to the observations of the common people, but it cannot be said either that a different general attitude would have been preferable. And a general attitude of some kind was necessary: some filtering of information had to be carried out in a rough and ready way, to reduce what remained to a manageable level.

Great changes have occurred between the time of this example and the present day. Science is now a clearly demarcated profession and the distinction between scientists and lay persons is more sharply defined and more significant than ever before. As far as the matter of who counts is concerned there is now something very close to a general rule: scientists count; others do not. The credibility which was once related to the word of a gentleman, and then to the standing of the philosopher and the virtuoso, now rests upon professional qualifications. General social standing and professional standing as a scientist have substantially separated, and it is the latter alone which is recognized as the legitimate basis for credibility in science. Admittedly, figures such as Emperor Hirohito of Japan are still elected to the Royal Society of London, but whatever this suggests about the easygoing ways of some natural scientists it does not really count against this point.

If it is true that scientists count and others do not, it is equally true that some scientists count more than others. Scientific standing is not used simply to specify who should be listened to; it is used as an elaborate indicator of how seriously they should be taken. The experience and the achievements of individual scientists define their professional reputation, their scientific eminence. And the more eminent a scientist the more readily and thoughtfully he is listened to. An oft-cited anecdote which illustrates this point well concerns Lord Rayleigh and a paper he offered for reading at a meeting of the British Association in 1888.[9] Rayleigh was one of the most eminent physicists of his day, but somehow in the process of submission his name had become detached from his paper, and a referee, commenting upon the anonymous draft, dismissed it as 'nothing but nonsense'. Apparently, when Rayleigh's name was eventually restored to the draft its quality was thereby so improved that it was accepted for reading after all.

Precisely because everything they say is taken seriously and evaluated with proper thoroughness, eminent scientists have an immensely important role in the general dissemination of scientific knowledge; they are able to act as sponsors of important new ideas and innovations and to use their standing to secure a fair hearing for them. Conversely, unknown young researchers with what they regard as important contributions, must expect some difficulty in gaining a hearing, and must often seek the ear of established and better known figures to champion their cause. This is particularly the case when the new contributions are unorthodox and at variance with existing ideas, so that on the face of it they very easily seem misguided or even utterly worthless.

A recent history of the early years of quantum theory has noted how slow, halting and ambivalent its early reception was in the physics community despite the very high level of the arguments which were used to support it.[10] But in 1908 Lorentz, a universally respected 'elder statesman' of physics, someone whose judgement and insight were everywhere admired and trusted, gave his backing to the theory in a widely publicized lecture. And from this point the diffusion and acceptance of the theory proceeded with great rapidity. Lorentz counted, and the results were clearly beneficial. But needless to say, any number of negative illustrations of the workings of the system are available to set against examples of this kind. Abel, a young and unknown Scandinavian mathematician sent his immensely important results to the world famous Gauss, but the paper containing them was found uncut and hence unread at Gauss's death. Gregor Mendel's epoch making memoir on heredity was similarly found amongst the papers of Charles Darwin. Abel did not count; neither did Mendel. And neither Gauss nor Darwin, both of whom did, chanced to pass beyond the unknown names upon the title pages into the substance of the papers they had received.

It is because of possibilities of this last kind, of course, that the injunction to look at the arguments and not at their author has so much point and relevance in the context of science. In the last analysis, it makes no difference to scientists who says what: all that matters is that it is well said. Scientists take no pleasure in overlooking work of the calibre of that of Mendel or Abel. But it is simply not possible to translate good intentions into improved practice. If either Gauss or Darwin had opened, read and seriously assessed every note and paper which came their way they would have been swamped with information and quickly rendered incapable of operating as competent assessors of anything. Here is the familiar dilemma again: science has to deal with the impossible task of avoiding information overload whilst overlooking no testimony of real significance.

Needless to say, as science has grown the dilemma has become more and more acute; more and more information is produced every year, more and more claims and testimonies have to be filtered and evaluated. The task is only manageable at all because as it has grown science has differentiated. There has been a massive specialization, an intense division of intellectual labour. If quality control in science is still a single system, it is a system made up of a myriad of sub-systems and sub-sub-systems, the links between which are in many cases extremely tenuous. Particular individual scientists, even quite eminent ones, find that their claims and judgements are taken seriously over a narrower and narrower range of topics: on any given topic those who count are all members of the relevant discipline, or speciality, or some even smaller *ad hoc* group recognized as possessing the relevant experience and expertise. And such narrow definitions of scientists' credibility are thoroughly justified by the narrowness of the competences of the scientists themselves, the very limited range of themes and issues to which they are content to devote their lives, the very narrow repertoire of tools and data with which they are content to work.

An amusing illustration of the extent of the intellectual division of labour in physics is provided by an episode which occurred at the very beginning of the century, when specialization in science was far less advanced than it is today. At this point many of Europe's finest theoretical physicists were interested in the way that solid bodies emitted heat, and in particular in how adequately classical theories and non-classical (quantum) theories of radiation described heat emission. The theoretical physicists spent an immense amount of time comparing these competing theories and assessing their relative merits. But they apparently failed to realize, until it was rather forcefully put to them, that the routine findings of experimental physicists had an important bearing on the matter. According to the indications of experiment, if the classical theory were correct then most solid bodies should glow in the dark at room temperature.[11]

No doubt a network of separate, specialized quality control systems is entirely appropriate given the vastness of modern science and the great range of theories and competences which it embodies. But inevitably it must operate in a less than perfect way. It must be expected to hinder the flow of information between disciplines and specialities, and perhaps to hold back innovations which draw upon the resources of more than one field. It is sometimes said, for example, that the introduction of important modern techniques of chemical analysis was needlessly delayed because they originated in the context of biology: if biologists had counted more with chemists then chromatography, and the analytical use of enzymes, both invaluable resources in modern chemistry, might have been developed and exploited much more quickly and effectively.

Arguments of this kind are difficult to evaluate of course, since we only have one history and we can never be certain what would have happened if this or that had been different. But the general point about the design of quality control systems must surely remain valid. As we seek increased reliability by using only the assessments of specialists, so we must risk losing, or delaying, the acceptance of important information. Needless to say, if this is the price to be paid for reliable knowledge, then it is worth paying.

WHAT COUNTS?

In 1969 a journal article appeared describing a somewhat surprising experimental finding.[12] According to the authors, when two beakers of water, of different temperatures but otherwise identical, were placed in a refrigerator and allowed to cool, ice appeared first in the beaker which started at the higher temperature. For example, 'If you take two beakers with equal volumes of water, one at 35°C and the other at 100°C, and put them into a refrigerator, the one that started at 100°C freezes first.' Moreover, for a given beaker of water, the higher the initial temperature the more quickly ice was observed to form within it upon its being placed in the refrigerator. On the face of it, the higher the temperature of water in given conditions the faster it freezes.

The authors of the article, E. B. Mpemba who initially noted the 'Mpemba effect' and D. G. Osborne who studied it more systematically at the University of Dar es Salaam, were clearly puzzled by the results of their work and felt that they demanded some kind of explanation. I suspect that most readers would agree with this, just as I do myself. (I suggest that the reader takes up pen and paper at this point, and spends a few moments working out what to make of the result and how it might be explained. It is useful to do this before reviewing the suggestions set down in table 2.1 (p. 60)).

Why then do the results seem so implausible, so distinctly strange and in need of explanation? For nearly all of us, it is not a matter of immediate direct experience: few of us will have the relevant experience of beakers, water and refrigerators. Nonetheless the results go against all our common sense intuitions. For a beaker at 100°C to freeze more quickly than one at 35°C it must cool from 100°C to 35°C and then from 35°C to 0°C more quickly than the other beaker does the second operation alone. The hotter beaker must do all that the cooler does and more, yet in less time, which just seems wrong. This conviction is reinforced for most people by their memories of school physics teaching. Newton's law of cooling states that the rate of loss of heat of a body is proportional to its excess temperature relative to its surroundings: it indicates that identical beakers at the same temperature in identical surroundings should take the same time to cool. (In school science Newton's law is, or was, often verified using the cooling of beakers of water.) Thus, on the face of it, if the 'Mpemba effect' were to exist one currently taught and accepted law of physics would have to be false. In fact, as anyone with more than a little knowledge of physics will immediately realize, it is not just one law of physics which is incompatible with the most natural way of perceiving and describing the 'Mpemba effect'. If, in otherwise identical circumstances, heat rushes across shallow temperature gradients with greater enthusiasm than it rushes across large ones, then quite a lot of physical theory starts to look a little sick.

Table 2.1

	Possible interpretations of the 'Mpemba effect' (warm water freezes faster than cool)
1	A practical joke.
2	An erroneous set of observations, perhaps due to faulty equipment or confusion of materials. Results not replicable.
3	Result of hot vessel melting into ice, and making better contact with cold refrigerator shelf.
4	Result of heating evaporating water, so that less is left to cool.
5	Something to do with supercooling.
6	Something to do with convection currents in the water.
7	Something to do with convection currents in the surrounding air.
8	Newton's law of cooling false.
9	Current thermodynamics and physical theories of heat false.
10	Modern scientific world view seriously defective.

The results seem peculiar, therefore, because they clash with our existing common sense expectations, which are in turn consistent with and supported by our awareness of the accepted knowledge of physics. How then might the results be explained? One possibility is simply to dismiss them as spurious. Perhaps they are a spoof or a practical joke. Or perhaps they are simply the result of some outrageously incompetent experiments, using faulty equipment, or neglecting elementary controls and precautions. A second possibility is that the observations were genuine and competently performed, but that they are in fact compatible with existing physical knowledge. If this is so, then it is necessary to look more deeply into the experiment and to call into question the way it was initially described. The experiment causes problems if its results were produced by *otherwise identical* beakers in *identical* circumstances. So far, this has simply been assumed. But should it have been?

There are all sorts of ways in which, even in a very carefully performed experiment, significant *differences* might exist between the hot and cold beaker. Evaporation is more rapid from a hot beaker. A hot beaker may melt through surface ice and make better thermal contact with the cold refrigerator shelf. A hot beaker will set up stronger convection currents, both within and around itself, than a cold beaker. Heat may very slightly alter the composition of the water by causing it to dissolve more foreign matter or by encouraging chemical reactions, and this may perhaps decrease its tendency to supercool. There are a number of further possibilities. Although none of them look all that promising, they do stand as potential ways of keeping the experimental results and keeping existing knowledge intact as well. Most of us, I suspect, would seek explanations of this kind (3 to 7 in table 2.1) once we were convinced of the good faith and basic competence of the experiments. Finally, the results are explicable as they stand, as events contrary to the indications of the currently accepted knowledge of physics: all that is necessary is to discard that knowledge as falsified and there is no problem. It is, after all, no adverse reflection on scientists if they got things wrong in the past, before the work was done with water and refrigerators: the scientists were not to know that these new results would turn up. And since science is tentative and provisional, it would be simple enough to adjust to new experience by setting existing knowledge aside. Nonetheless, however easy this may be, I do not imagine for one moment that I, or anyone else, would seriously contemplate taking such a step. I list 9 and 10 in table 2.1 purely as formal possibilities not as realistic ones.

I do not know the current state of play with regard to the 'Mpemba effect'. Certainly, it has not come to be accepted as part of the main body of scientific knowledge; indeed it has never been allowed a genuine chance to do so. It will come as no surprise to hear that the effect was reported in *Physics Education,* not in *Physical Review Letters* or the *Proceedings of the Royal Society,* and that it has never counted as a fully accredited research report but only as interesting material for student exercises and projects. However, that is not to say that the finding has been refuted, or explained away. To my knowledge, this has not occurred. Whether or not the effect is genuine is, however, of no consequence as far as my present purposes are concerned. My sole reason for discussing the effect is to draw attention to how it is thought about. It very effectively illustrates *the crucial role of existing knowledge* in assimilating and describing new findings. The puzzling character of the 'Mpemba effect' derives entirely from its relationship with existing knowledge – from the fact that *prima facie* it is incompatible with that knowledge. And the 'explanation' of the 'Mpemba effect' is largely a matter of making it compatible with existing knowledge – of redescribing it in a way which no longer conflicts with that knowledge.

The role of existing knowledge in the assimilation of new experience and the creation of new knowledge is basic and fundamental. The issues it raises go far beyond what I am able to allude to here. Scientific research is just the use of existing knowledge to extend and modify that very same knowledge. The simple example described above, although not perhaps to be taken seriously in itself, provides a useful illustration of the general theme. Indeed, now that the purpose of the example has been made clear, it is worth returning to it for one last time.

There are two mental exercises, or thought experiments, which are worth trying out using the example. First, try to describe the experiment, make sense of it, advance possible explanations of it, without relying upon existing taken-for-granted knowledge. By actually attempting to carry out any of these tasks, and experiencing the difficulties involved directly, the importance of existing knowledge can be appreciated vividly and immediately, even in this case where the experimental situation is so extremely simple. The exercise is an excellent route to understanding that we never encounter empirical experience naively and directly, but always in a way which is mediated by our existing knowledge.

As a second thought experiment, review all the accounts you can readily think of for what is going on as the water cools, and try to get a sense of the difficulty of deciding between the various alternatives. Notice also the additional accounts which you could put together, given more time to think about beaker wall conductivities, surface contact effects, viscosity, and so forth. What I am asking is that you seek a sense of the devilish complexity of the empirical situation here, and the difficulty (should I say the impossibility?) of giving a full, exhaustive account of all its facets and features. Why do this? The point is that scientific theories are generally believed to have straightforward logical implications for particular empirical situations. It is these straightforward implications which are thought to enable scientists to make empirical predictions using their existing knowledge, or to test their existing knowledge against empirical experience. But in this situation it should be terribly clear that the implications of existing scientific knowledge are anything but straightforward. There is indeed no clear way of deciding beyond all doubt how best to describe the situation in terms of existing knowledge. Yet the situation is simplicity itself: two beakers of water and a refrigerator. And the relevant existing knowledge is routine and well worked out. If things are unclear here, surely they must always be unclear.

This is indeed precisely what I take to be the point of the second thought experiment. It shows that although we always make sense of actual situations using existing knowledge, the sheer complexity of actual situations means that no particular way of so making sense of them can ever be put *completely* beyond doubt. This point is a difficult one, and not one which I shall analyse further in this book. It is, however, worth bearing in mind because of its relevance to later themes. What it amounts to is that people may trust and respect science generally, or any field of science, and yet legitimately doubt how it is actually being applied in any specific situation: if there are problems about the proper scientific account to be given of two beakers of water in a refrigerator, so similarly will there be problems about the proper account of a nuclear reactor, or an eco-system. It is very much to their credit that natural scientists do not claim certainty for their knowledge, or for any particular applications they make of it, but there is a need to beware of other people who continually tend to make such claims on science's behalf.

Let me, however, conclude this digression and move back to my central theme. In one way or another, people are necessarily reliant upon their existing knowledge in assimilating and making sense of new experience. This is as true of scientists as of anyone else. Scientists make use of what already counts as knowledge in deciding what should come to count as new knowledge. They made use of existing knowledge in deciding that no version of the 'Mpemba effect' should count as new knowledge. New knowledge claims are always assessed in ways which rely heavily upon existing, accepted knowledge. And this existing, accepted knowledge is not the personal knowledge of the individual, distilled from his or her individual experience. It is the inherited and shared knowledge of the community; in the case of a scientist it is the shared knowledge of a given scientific community. Individuals typically acquire this knowledge from teacher or texts, much of it in the course of their training, and they tend to trust it and to take its validity for granted. Very few physicists, for example, feel any need to give the laws of heat flow a personal check-over. Indeed, if they were to obtain results which were at variance with the laws, it would probably be the results which they rejected.

Earlier in the chapter, I spoke of the prolonged and intensive training which all scientists receive, and of its importance in aligning their thinking and increasing their collective effectiveness in the generation of new knowledge. It is now possible to take this point further. It is training that largely imparts existing, accepted knowledge. And it will now be clear that accepted knowledge is not merely a stockpile which is added to by researchers who proceed purely on the basis of observation and a few rules of scientific method. Accepted knowledge is a key resource all the time in doing research. It conditions the judgement of scientists profoundly and pervasively at every point. Quite simply their systematic use of their existing knowledge is what makes natural scientists so very much more effective in the activity of research than any group of outsiders could hope to be.

How are we to think about this role for existing knowledge in science? Clearly, if scientists could but start with a small body of rock solid truths the system would be marvellously effective. Truth is presumably a coherent whole: anything which contradicts a truth must be false. So once a few truths are established as existing knowledge, anything further which clashes with them can at once be discarded. A massive simplification can be achieved in the practice of research. Many people do indeed take the existing knowledge of science in just this way, and use it as a set of truths against which quickly to check and test other knowledge claims. They say, for example, that reports of extra-sensory perception and parapsychological phenomena must be false because they conflict with the (true?) laws of physics. Or they make the same assertion of the Biblical accounts of Christ's miracles, and of the account of the creation in Genesis. There is indeed a vast range of knowledge claims, advanced by religious sects, pseudo-scientific fields, cranks and eccentrics, which are confidently dismissed in many quarters purely on the grounds of their incompatibility with the 'truths' of science. If only it were that simple.

This book is not about the nature of scientific knowledge and its standing as an account of physical reality. These fascinating subjects need a book to themselves, and I have deliberately done my best to avoid discussing them here. However, a certain amount of what I have to say does rely upon the generally accepted view that the knowledge of science is provisional and uncertain, and I feel obliged to say just a very few words about why this view is generally accepted and why it is right. There are a number of different ways of talking about the problem and coming to understand it; the different schools of philosphy of science all have their favoured ways of speaking of it. But perhaps the simplest approach is to emphasize the theoretical nature of science. The existing knowledge of science comprises not a direct reflection of the real world but a theoretical interpretation of that world. Theories are invented by us, and used to describe and interpret the world, or rather the observations and experimental findings in terms of which it is known. But it is a straightforward logical point that any finite set of observations and findings is compatible with any number of theories. This is just like saying that any number of curves can be drawn through any finite number of points plotted on a graph. Accordingly, no set of observations and results accumulated by scientists can ever suffice conclusively to establish a theory, since however hard scientists work the data they generate are always finite.

One might accept this argument and yet wonder whether too much is not being made of it. Is there not, after all, a factual component in existing knowledge, as well as a theoretical one, and cannot we at least rely upon the facts? The difficulty here is that scientific facts and scientific theories are inextricably mixed up with each other; there is no independence of fact and theory. One way of understanding this is to think of the actual business of obtaining a fact by measurement or observation. To know that the fact is indeed there, one needs knowledge of one's equipment or apparatus; but the only way to derive this knowledge is from a theoretical understanding of how the apparatus operates. If the theory of the apparatus changes so do the reported facts which it is said to have measured.

This may seem a very fussy kind of argument, but it actually highlights something which scientists themselves find a continuing problem as they carry out their research. The history of microscopy, for example, is a continuing struggle reliably to separate facts about phenomena from artefacts of the apparatus and the method of observation itself. Past difficulties are reflected in current training: it is important, for example, that biologists know how to recognize lens aberrations and distortions in their many shapes and forms, and that they are not mistaken for genuine phenomena. Present difficulties are reflected in technical debates amongst researchers. It was recently proposed, for example, that the majority of the observations so far made on biological materials with electron microscopes are spurious and have merely identified patterns generated in the preparation of the materials. This suggestion has now been largely rejected, after due consideration by those in the relevant fields, but that practising scientists found themselves obliged to formulate it, and give it extensive consideration, well illustrates the point that 'the facts' do not stand as indisputable points of reference for natural scientists.

The problematic nature of scientific facts and the impossibility of disentangling them from scientific theory is something which gives rise to difficulties even at the most fundamental level of scientific research. In elementary particle physics recent attempts to identify quarks, particles with a fractional electrical charge, have produced an interesting controversy about 'the facts'. An elaborate experiment performed in the United States with highly sophisticated apparatus and by experienced and respected investigators, has led to the claim that a fractionally charged particle has indeed been identified, and isolated on a small sphere of Niobium. But an alternative view of this same experiment is that the 'data' it has yielded merely reflect spurious effects present in the apparatus. Potentially, this is an experiment of great importance, since it could count as a check upon important aspects of physical theory. It was theoretical development which initially pointed to the possible existence of quarks and inspired the experiment; but later theoretical developments have suggested that although quarks should exist, the way that they attract each other should make it impossible for them to split off from each other and exist in isolation. Clearly then, the matter of whether there really is, or was, a quark sitting on a Niobium sphere in the United States laboratory is of considerable interest.

Naturally, it is likely that attempts will be made to develop and improve the quark identifying experiment, and make its results that much more compelling. But the experiment is perhaps even more complicated than that involving two beakers of water and a refrigerator described earlier, and it is not easy to imagine how it might be refined so that the real existence of the quark is put beyond all doubt. Given this, it is possible that theoretical developments may decide how the experiment is interpreted in the long term. If the most satisfactory overall theoretical picture to emerge in the next few years is one wherein quarks just cannot wander about by themselves, then this will indicate that there was not really a quark on the Niobium sphere after all. Physicists will then be confident that the apparatus has a bug in it. Theory will have helped the physicists to work out what the facts are.

Science is theoretical knowledge. And it is theoretical through and through, not just in part. Scientific knowledge is theories which we or our predecessors have invented and which we remain content to use for the time being as the basis of our understanding of nature. This is a very commonplace account of science, and it does not indicate a need for any major alteration in the way we habitually think about it. Certainly it does not imply that we should cease to trust and use scientific knowledge: on the contrary, scientific knowledge is just the knowledge we have found most trustworthy in use. But bearing this very commonplace account in mind can serve to keep us alert against claims and arguments which make a fetish of science – which assume that our scientific knowledge is permanently valid and is justified wholly and entirely by its correspondence to reality. It is remarkable how widely such views are still entertained. After all, the majority of all the theories which scientists have ever put forward have been rejected as false or misconceived, and the majority of the findings which they have reported have been forgotten. Scientific knowledge has an extremely short lifetime. The knowledge routinely accepted and used in any scientific field is on the whole extraordinarily recent: scarcely any fields make use of materials more than a few decades old, and such older material as is used is very rarely accepted just as it stands. Yet because we place such trust in it now, many people have difficulty in seeing that our present knowledge is likely to be treated in three or four generations much as we ourselves treat the knowledge of three or four generations ago. The need to trust engenders the desire for certainty; and what people desire to believe many of them actually end up believing.

The existing, accepted knowledge of science, passed on in the course of training in all the various scientific fields and disciplines, is very effectively used in the practice of scientific research. Scientists rightly employ this knowledge as a criterion of what counts as good work. They quite justifiably operate the systems of recognition and reward in their various fields in ways which tend to take this knowledge for granted, and which require some kind of a special case to be made if any significant component of it is to be called into question. And they quite properly make a distinction between this special scientific knowledge, accepted in their field, and the less trustworthy knowledge established and taken for granted in everyday life. Nonetheless, the accepted knowledge of a scientific field cannot be thought of as a set of fixed truths: it is indeed continually changed as it is used. It is a developing interpretation of the world, or a part of the world, not a reflection of the world: it is not guaranteed and secured solely by reality itself.

One very important consequence of this is that different, potentially *conflicting* bodies of accepted knowledge may grow up in different scientific disciplines. And scientists may conflict in their judgements, in their views of what counts in research, because they operate with reference to different bodies of accepted knowledge. The unity of science may, to this extent, be illusory. Naturally, when knowledge in different fields of science is seen to clash, scientists will turn their attention to the matter and work out some way of re-establishing consistency. But it is as well to be aware that although all the different fields of science claim to be describing the same physical reality there is never any *advance* guarantee that all the various alternative descriptions they make available will be consistent with each other.

Perhaps the best known case where substantial bodies of accepted scientific knowledge have been set into conflict with each other is that of the nineteenth-century controversy concerning the age of the earth.[13] This controversy, which began around 1860 and did not entirely subside until well after the First World War, involved many eminent scientists in the English-speaking world, and concerned the accepted knowledge of geology, biology and natural history on the one hand, and physics and astronomy on the other. Geology had successfully established itself as a respectable and increasingly professionalized scientific activity in the first half of the nineteenth century. It achieved considerable success in accounting for the laying-down and shaping of the earth's rocks in entirely matter of fact terms, which from a scientist's viewpoint contrasted very favourably with earlier explanations in terms of divine interventions and the catastrophic flood reported in the Bible. Many geologists made a point of explaining past geological events wholly and entirely in terms of the natural processes and changes which were readily apparent to contemporary observers: they adopted this principle of uniformity of explanation as a properly 'scientific' one, and thereby distanced themselves from the earlier theological speculations, and the imaginative excesses of less reputable investigators. And they found they could make considerable headway with their new approach provided only that they accepted an extremely generous time-scale for the history of the earth. Uniformitarian geology made use of 'unlimited drafts of time' in order to explain geological phenomena in a matter of fact, 'properly scientific' way.

In this respect geology was at one with the new theory of speciation by natural selection introduced by Charles Darwin in 1859 in *the Origin of Species.* Darwin recognized that for natural selection to generate the variety of species known to natural historians vast time periods were required. He took the work of geologists as evidence that such time periods were available, and rejoiced in the consistency between the requirements of biology and geology. In his book he referred to the writings of the uniformitarian geologist Charles Lyell for support for his required time-scale; and to illustrate what geological reasoning suggested he noted that the mere process of the denudation of the Weald, i.e. the eroding away of the valley between the North and the South Downs, must have taken, projecting from its estimated current rate of erosion, something like 300 million years.

Unfortunately, the availability of time periods of this magnitude was rapidly called into question by Britain's leading natural philosopher William Thomson, Lord Kelvin, who extrapolated from physical principles, and the currently accepted knowledge of physics, chemistry and astronomy, to set an upper limit upon the age of the earth. One of Kelvin's main lines of argument involved assuming that the sun had to continue to supply heat at roughly its present rate over all the long period of uniform geological time. What was the supply for this heat? Kelvin sought to uncover every likely source. Calculation showed that chemical sources of energy could plausibly account for but a short period of energy dissipation. The most promising energy source was in fact gravity, that is the potential energy changes involved in bodies moving in gravitational fields: meteors falling into the sun, for example, might contribute very substantial amounts of energy to that body and maintain its temperature for a long period. Having extensively reviewed what he took to be all the possibilities of this kind, Kelvin set an upper limit on the age of the earth of 100 million years.

The resulting controversies were long and very complicated. There can be little doubt that moral and religious commitments were to some extent implicated, particularly in so far as attitudes to Darwin's theory were concerned. But even at a narrowly technical level the debate is difficult to summarize. Starting assumptions were reviewed and re-evaluated. Existing knowledge was reflected upon. New contributions were made, some of which ameliorated, others of which intensified the clash of results. A significant proportion of geologists capitulated to the physical argument, or even in some cases welcomed it. Nonetheless it is probably fair to say that for a long period following Kelvin's work scientists were aware of an uncomfortable incongruity between the doctrines of physics and those established in geology and biology. At least the incongruity was uncomfortable for the supporters of the theory of natural selection and a uniformitarian style of geology. Although these men had confidence in their interpretation of the geological and biological evidence, and regarded work in their fields as providing ever-increasing support for that interpretation, they were impressed by arguments based on physics and recognized them as a worrying threat to their own convictions. Amongst physicists and natural philosophers on the other hand the incongruity was experienced differently: nobody seems to have perceived the evidence of geology and biology as disconfirmation of the laws of physics. It is interesting to ask what is at the heart of this asymmetry. Does physics always have better-quality evidence and superior theory to biology and geology, or is it merely a matter of physics standing at the top of a conventional hierarchy of the sciences?

As it happens no significant revisions to the shaky doctrines of geology and evolutionary biology were required to resolve this controversy. It was the accepted knowledge employed by Kelvin which was radically revised, as the imposing edifice of late nineteenth-century physics was shaken by a whole series of revolutionary disruptions. Radioactive dating techniques vindicated the demands for vast tracts of time made by the geologists and biologists. And new sources of energy residing in the nucleus of the atom allowed physics correspondingly to extend its estimate of the age of the earth.

Since the time of this controversy the knowledge of the natural sciences has grown and differentiated on a massive scale. An immense amount of inherited knowledge now exists in a highly fragmented form, with the various fragments maintained and passed on in the many distinct and separate scientific disciplines. It is pointless to ask how far all the fragments constitute a single, consistent, coherent whole. One cannot compare segments of knowledge with each other abstractly and try to fit them together like the pieces of a jigsaw puzzle. The only worthwhile answer to the question is that which scientists themselves will provide as they carry out their research: any incompatibility in the different areas of presently accepted knowledge will be revealed in practice.

There is, however, one final question which must be asked about these fragments of inherited, accepted knowledge. I have said that they do not simply reflect reality, and that they are therefore not sustained and justified simply by their correspondence with reality. Yet natural scientists typically have immense confidence in the accepted knowledge of their field. Even before they have actually made use of their knowledge, at the end of their training and the beginning of their careers, they are very well aware that what they know is thoroughly worth knowing. How is this so? If its truth cannot be made transparently manifest, what makes scientific knowledge convincing?

Much of the answer must be looked for in the process of scientific training itself, at the higher levels which qualify students for professional scientific work. This professional training does not even seriously attempt to justify and validate the knowledge it conveys: it very much presumes that the knowledge will simply be absorbed and accepted, as indeed it is. One reason for this ready acceptance is that students are acquiring primarily not a world picture or a theoretical scheme, but a craft. As I described in the previous chapter, they are gaining scientific competences and skills in a process rather like an apprenticeship. Much of the knowledge they acquire is procedural, and the value of it can be seen directly in terms of the use of the procedures. Other, more explicit and verbal knowledge is acquired in association with procedure, as a natural accessory to it. The situation can be thought of as analogous to the learning of music, where a whole range of technical skills is acquired, and formal verbal knowledge is also acquired along with the skills as part of the package. It may or may not be relevant that both in the case of the professional scientist and the professional musician the package is seen as the necessary equipment for a future occupation.

There is, however, a further reason why natural scientific knowledge is so readily accepted. This is that teachers, having themselves agreed what in their field is reliable and important, and having constructed a curriculum accordingly, generally take care to invest that curriculum with *authority*. Typically, the curriculum is embedded in a textbook or books which set out the key components of accepted doctrine as *the* correct way of working at things, and which present them in the best possible light. Texts are expressly designed to create conviction: they usually present one interpretation only, play down any problems and uncertainties currently associated with the interpretation, and occasionally present a heavily idealized picture of its historical development wherein it seems to be the only tenable account of accumulating data. This style of presentation is usually reinforced by teachers themselves, in lectures and small group teaching. Science students are not on the whole encouraged to treat their literature as philosophy students, for example, treat theirs. Philosophy students are expected to pull arguments to pieces; look for weaknesses, lacunae in reasoning; set off one interpretation against another, one writer against another. There tends to be only a nominal gesture in this direction within a scientific training. The main point is to assimilate the material and become competent in its use, not to become aware of the omnipresence of uncertainty and thereby become at risk of demoralization. Scientific training has a quasi-dogmatic character. This is not always easily acknowledged by scientists, who would naturally prefer beliefs to be accepted purely on the basis of reason, and not at all on the basis of authority and subtle compulsion. But the dogmatic character of training in most contexts is evident enough. In a certain sense it exists by necessity. Science is an occupation, and if it is to continue to exist as an occupation the relevant knowledge simply has to be learned. If science offered the only account of the natural world tolerable to human reason, there would be no difficulty. But this is not the case. Awareness of reality cannot compel the acceptance of scientific knowledge. So what the world cannot do, the scientific community itself takes on, and does in its stead.

3

Authority

AN EXPERIMENT

A good part of the science of social psychology is concerned with the extent to which individuals are influenced in their behaviour by others around them. Social psychology studies our tendencies to conformity and to obedience, and the extent of their consequences. Its literature makes depressing reading: it is a long catalogue of values, beliefs and actions being produced and adjusted with a mind to their social acceptability and the requirements of authority.

I want briefly to describe one of the classic experiments in social psychology. It is just one of a series of experiments, all with the same basic form, which were set in train by Stanley Milgram nearly 30 years ago in an attempt to study the extent to which people will obey authority.[1] These experiments and their results are now justly famous, and it may be that many readers of this book will already know of them. Their interest is such, however, that there is no danger of undue attention being given to them. Everyone should know of Milgram's work, and natural scientists in particular should all be aware of its findings.

In seeking subjects for his experiments Milgram tried to recruit a typical cross-section of the adult community, and indeed over the years of his work he did not find any massive variations in his results according to the subjects he chose: his basic findings seemed to apply to men and women, old and young, rich and poor. It is worth noting, however, that Milgram did his work in the eastern United States in the early 1960s, and that in the particular variant of his experiment I shall be describing the subjects were males aged between 20 and 50.

Milgram obtained subjects by advertising for assistants to participate in a scientific experiment. Selected subjects arrived at Yale University where they were met, paid a small sum for their services, and taken to the psychology laboratory. Here, they were introduced to the scientist carrying out the experiment, the experimenter, and given details of what was to be done and what their role in it was to be.

The experiment was described as a study of the effects of punishment on learning. The subject was to act as a teacher who was to administer a memory test to another participant, a learner. The teacher was to administer graduated punishments to the learner: a punishment of increased severity was to be administered on each successive occasion that the learner failed to provide the correct test answer. The learner was placed in a room adjacent to that of the teacher, strapped into a chair and connected so that he was obliged to receive electric shocks as punishments. The teacher was provided with the means of administering the shocks, an array of 30 switches which connected voltages ranging from zero to 450, at intervals of 15 volts. A shock of 45 volts was given to the teacher to demonstrate the distinctly unpleasant sensations it produced, and to show that the equipment was functioning correctly. The teacher was then left to administer increasing levels of shock to the learner whenever incorrect responses were made.

Needless to say, the experiment was actually designed to investigate, not the effect of punishment on memory, but the extent to which people would be willing to punish at the behest of others. The learner, the victim of the supposed punishments, in fact received no shocks at all and was the knowing accomplice of the experimenter. The teacher was the only true subject, the sole focus of scientific curiosity; the question was how far he would move along the row of switches as the learner gave more and more incorrect answers to the memory test.

Within the immediate situation, the only clear incentive to proceed with the administration of more and more severe shocks came from the experimenter. When turned to by a dubious subject he would state that the experiment should proceed. He would indicate this by using a restricted range of expressions of authority and command, set down as part of the highly standardized conditions under which the experiment was conducted. The strongest of these expressions was: 'you have no other choice, you must go on'. And the expressions were, of course, uttered in a standardized way, without emotion or threatening gesture.

What were the disincentives to the subject's proceeding to the end of the line of switches, and eventually administering shocks of 450 volts? They were of two kinds. First, upon the shock generator itself, the voltages associated with the various switches were clearly marked, so that the subject was clear that the third of 30 switches had produced his own disagreeable experience; and the verbal messages 'Extreme Shock' and 'Danger: Severe Shock' were also displayed at successive points later in the sequence. Secondly, the learner in the adjacent room was heard producing increasing signs of distress. At 75 volts, grunts and moans were audible; at 150 volts, cries for release; at 180 volts, assertions that the pain was unbearable; at 300 volts, refusals to participate further and demands for release; from 330 volts upwards screaming followed every throw of the switch.

How far then would the typical subject proceed along this formidable obstacle course? One might think not very far, if indeed people were willing to begin in the first place. Milgram himself put the question to audiences when he described his work: he would describe his experiment and then ask for predictions of his results before actually divulging them. There was broad agreement that scarcely anyone would complete the experiment and that most subjects would have dropped out well before the half-way stage. When Milgram solicited 'expert' views, those of the psychiatrists in a United States Medical School, the average response was that less than 20 per cent of the subjects would continue to the half-way point, and that less than one subject in a thousand would throw the final switch and complete the experiment.

The 'expert' predictions were in fact even more astray than those of 'unqualified' persons. What Milgram had actually found was that almost 80 per cent of subjects had continued to the half-way point, and that more than 60 per cent had thrown the final switch.

These are indeed extraordinary results. It is important to dwell upon them for a moment, turning them over in the mind, lest the imagination should seize some opportunity to diminish them, or avoid facing their full significance. Some 60 per cent of subjects, 60 per cent of a quite normal representative groups of American men, had completed the experiment. Each of them had, as it appeared to them, inflicted 30 electric shocks on another person, all technical assaults punishable at law. Each had ignored or overridden a visual warning of danger, and successive protests and entreaties from the recipient. Seven times in succession each one had elicited agonized screaming by throwing a switch. All this had been done voluntarily, without threat or physical coercion, with nothing to prevent the subject ceasing to act and walking out of the situation. All that had been required to enforce the implementation of the experiment had been the verbal instructions or orders of the scientist/experimenter.

That this degree of obedience was encountered came as a surprise to Milgram himself. In his earliest experiments, with which he expected to identify only a small number of unusually obedient persons, he had incorporated no aural feedback; but 'In the absence of protests from the learner, virtually all subjects, once commanded, went blithely to the end of the board.' Milgram had deliberately to introduce protests and screams on the part of the victim in order to induce even a minority of subjects to disobey the experimenter. He did this because initially he was interested in ascertaining whether any interesting differences existed between obedient and less obedient persons. In the end, however, this aspect of Milgram's work proved far less interesting than did its implications concerning the behaviour of people generally.

The Second World War, and in particular the systematic atrocities perpetrated by the Germans in the course of it, were an important formative influence as Milgram created his programme of experiments. In the course of the war people rose every morning and set out to their duties in concentration camps and extermination camps. Others participated day after day in the development and production of the associated equipment and artefacts: the gas chambers, the lethal substances. The business of killing large numbers of helpless innocent people as rapidly as possible became for many part of the routine business of everyday life; the stench of burning bodies became a constituent part of the work environment; the processing of these bodies became a component of the economy. Once the mind had begun to face this, and cope with this, it was bound to ask: how was it possible? How could people live a life which did not just impinge upon such practices but was such practices? How could such a life be endured, let alone sought after?

The beginnings of a possible answer were provided after the war by some of the protagonists themselves, as they sought to exculpate themselves from what had happened in the extermination camps. They had, they said, merely followed orders: the responsibility lay with their supervisors; they had simply done as they were told. Claims of this kind were never accepted as excuses after the war (that is, as far as enemies were concerned), but they were taken seriously as explanations. Perhaps a certain type of person existed who was particularly susceptible to authority, and who would indeed carry out practically any task, however offensive or appalling, on the command or at the behest of a legitimate authority. Perhaps the German camps had been able to recruit sufficient numbers of such people, with this kind of personality, and had been able to operate without difficulty as a result. If this were so, then the existence of such people was a fact of great social and political importance, and their identification was a task meriting urgent attention.

Milgram's experiments were designed to identify such people and indicate something of their numbers and their distribution. In their own terms of reference I suppose it can be said that in this they succeeded.

INTERPRETATION

Milgram's experiments are merely the most spectacular of a great number in social psychology which display the susceptibility of the individual to social pressures of various kinds. Milgram considered the actions of subjects: other workers have produced equally striking results concerning their expressed beliefs and values, and even their estimates of physical quantities such as length. Milgram considered the effects of authority: other workers have analysed the comparably strong effects of peers – other members of a group of which the individual is a member.

Milgram himself actually modified his own experiment to study the power of the opinions of peer group members. He placed his subject in a small committee charged with the operation of the electric shock machine, and arranged that committee members would oppose the instructions of the experimenter, so that the subject was torn between the recommendations of authority and those of peers. In these conditions the completion rate dropped from more than 60 per cent to a very low level indeed: the peer group largely cancelled out the influence of the experimenter. On the other hand, when the committee was made to support the instructions of the experimenter the completion rate shot up to over 90 per cent. This indication of the strong susceptibility of individuals to peer group pressure is consistent with the results of many other series of experiments.

In the main, however, Milgram's work was concerned with obedience to authority, not with simple conformity or with the emulation of the norms and practices of a group. This indeed is why I have introduced Milgram's work into this book. Milgram showed experimental subjects recognizing the authority of a *scientist* and following his orders. When Milgram arranged that the experimenter should present himself as a lay person, and deliver his orders, instructions and requests entirely in the manner of a layman, the completion rate in the experiment dropped to around 20 per cent. Clearly, a great part of the authority exercised by the experimenter derived from his being perceived as a scientist, as a figure qualified and competent to direct affairs in the circumstances of the laboratory.

In an ideal world, all scientists would know of Milgram's work. It is a reminder that they possess authority, and it vividly exemplifies the potentialities of that possession. It is not something to be made light of; it is a large matter. An awareness of Milgram's experiment, a lasting memory of an image of it, serves as a reminder of the size of the matter.

On the other hand, it would be quite wrong if this very much simplified presentation of Milgram were to be swallowed whole. Any detailed discussion of Milgram in social psychology would have to ask a great number of questions concerning how his results should be interpreted and what inferences might properly be made from them. Alternative interpretations would be set in opposition to each other, and further experiments would be suggested to help to decide between them. As a matter of fact comparatively few further experiments have been performed following Milgram's model: the electric shock machine with its 30 switches has not become part of the standard equipment of the social psychology laboratory. One reason for this is that the experiment raises important ethical problems: it is extremely stressful for subjects and imposes the stress through the use of a prolonged and systematic deception. Social psychology continually confronts fundamental ethical dilemmas; and over the years it has become more and more cautious wherever it draws near to them. What Milgram, after due consideration, felt it right to do in the particular circumstances of his time is no longer so readily undertaken. Nonetheless, Milgram's results remain a standard point of reference in social psychology, and the precise significance of his results continues to excite debate.

This is not a textbook of social psychology. I shall make no attempt to describe the many points of difficulty involved in the interpretation of Milgram's work, nor indeed could I if I wished to do so. But I do feel obliged to make two basic points which should always be borne in mind in connection with work like that of Milgram.

First, it would be wrong simply to accept his findings as 'facts about obedience to authority'. No experimental findings ever interpret themselves, and no particular interpretation placed upon them is ever devoid of problems. It is Milgram's *theory* that in his experiment subjects acted as they did out of obedience to authority. It is, I think, a good theory: Milgram gives good grounds for accepting it in preference to alternatives. But a theory it remains, and it cannot be accepted as conclusively established. Moreover, the very vocabulary of the theory raises a host of further questions. What precisely is the 'authority' of the experimenter? Is it based on a relation of trust, or one of domination? Is it a personal possession, as it were, or something associated with the experimenter's role? What would happen if the experimenter gave orders anxiously, or if he confided that he himself was merely acting under orders and knew little of the experiment, or if he were to wonder, as the victim cried out in pain, whether something might have gone wrong? Would any of these actions on the part of the experimenter effectively destroy his authority, and if so what would that imply about its basis? Then again, assuming that the authority of the experimenter is indeed the cause of obedience in the experiment, there is the question of how it operates, and in what range of circumstances it will remain effective. The progression through the sequence of switches step by step slowly evokes from the subject more and more severe actions: would these actions have been evoked directly and at once by the instructions from the experimenter, and if not why not? Similarly, the physical presence of the experimenter is a factor in the experiment, and it is known that as the experimenter is physically separated from the subject, the tendency to obedience weakens. Why should the tendency to obedience weaken to some extent in this fashion? And if it is due to a weakening of authority itself, with distance as it were, what does this imply about the nature of authority? There is a great range of questions of this kind which must be asked. For some of them plausible answers exist within the framework of our current understanding of human behaviour; for others there are no such answers. Thus, whereas it may be perfectly reasonable to speak of Milgram's subjects obeying authority, it is nonetheless just possible that we may come to decide that such a way of speaking is incorrect. And even if we continue to accept it as correct we must recognize that we are not fully clear as to what we mean by it.

Secondly, there is the matter of what the accepted interpretation of Milgram's results should properly imply for other people, at other times, in other circumstances. Let us suppose it correct that in the early 1960s a large number of American citizens completed the shock experiment, and thereby revealed their high susceptibility to the influence of a scientific authority. What does this tell us? Does it suggest that human nature makes us susceptible to authority? Or does it indicate that Americans raise their children to defer to authority? Or should we be careful to speak only of the *scientific* authority involved in Milgram's experiment? And then there is the matter of time: because some people were obedient at one point of time, can we infer that other people will be obedient at a later point of time, or even that the same people will be? How do we know for example that it was not the effects of wartime experience and the military form of life which so many people had to lead, which produced the high level of obedience noted by Milgram?

This is the deep problem of the *projectibility* of knowledge to which I have alluded in an earlier chapter. In its most general form it is simply the problem of induction itself: how can anything about the future be inferred from what has already passed? How do we know that electrons will behave next week as they behaved last week? How do we know that electrons we have not so far encountered are going to be just like the ones we have encountered? This is a problem which afflicts physics as much as social psychology, and one to which nobody has ever produced a satisfactory answer.

But in social psychology, as in the other human sciences, the problem has an additional aspect which does not have to be faced to the same extent in the context of physics. With electrons, our grounds for believing that they will go on behaving in the same old way may not be particularly secure, but at least we have no clear indications that they will not. With people, on other hand, we do have such indications: we know that human behaviour does indeed manifest massive variation from context to context, culture to culture, and that attempts to identify invariants in that behaviour have a very poor track record.

As far as a specific experimental result is concerned, we become confident of its general scope and significance if we can fit it with other related pieces of evidence and if for practical purposes it is replicable – particularly if it is replicable in varying conditions and circumstances. It is tempting to attribute very general significance to Milgram's experiment because it fits nicely with other experiments and findings in social psychology, and because it helps to make sense of so many diverse depressing episodes in human history – episodes which occur, indeed routinely occur, in a wide range of cultures. On the other hand these are also grounds for exercising caution in what we infer from the experiment.

To my knowledge there has been no recent repetition of Milgram's work. And even if social psychologists were willing, such a repetition would probably be impossible to execute, since the experiment is now so well known that a successful series of deceptions would not be feasible. But a closely related series of experiments has been repeated recently. In the 1950s S. E. Asch showed how estimates of length could be profoundly modified by social pressure. Individuals were placed in a group (of the experimenter's accomplices) and asked to estimate the length of a line. If the other members of the group all massively over-estimated the length, then the individual subject also would move strongly in that direction and likewise over-estimate the length. This finding was reliably reproduced by other experimenters through the 1950s and 1960s and was taken as good evidence of the tendency of individuals generally to conform to group opinion. But in 1980, in Britain, the experiment was repeated using 396 natural science students as subjects: only one subject made his estimate conform with that of the erroneous majority.[2] It would be wrong to make more of this 1980 experiment than of previous experiments. But it does highlight the limitations of our knowledge. We cannot say with confidence of any experiment in social psychology, that the next time it is done it will not produce results quite contrary to all those previously obtained until that point.

THE AUTHORITY OF SCIENCE

Striking as they are, Milgram's results have to be treated with caution and reserve, treated, that is, just like the results of all other experiments. And the fact that they concern human behaviour adds to the force and significance of the point. But I do not want to bury the results under a pile of reservations and qualifications, so that they can be discounted and forgotten. What Milgram described did occur, and his way of describing what occurred suffices for practical purposes. It seems reasonable to accept that in his experiments subjects were extraordinarily obedient to persons they believed to be scientists, and that they were obedient largely because those persons were believed to be scientists. And this must surely be taken as a token of the potency of the trust and authority which may be invested in scientists, of the readiness with which their definitions of situations may be accepted and their recommendations for action carried out. Moreover, given what we can see around us of the current standing of science and of general attitudes to it, it seems plausible to suggest that the extent to which science is trusted and its authority accepted today is entirely comparable to that at the time of Milgram's experiments.

Whatever else, Milgram's work should be taken as a reminder of what the possession of scientific authority involves, and a warning of what its potentialities can extend to.

How should a scientist respond to such a warning? One way is to see authority as something that science has acquired by accident, something that the public has imbued it with, and which scientists themselves have perhaps carelessly and unthinkingly accepted. From this point of view, the value of Milgram's work for scientists is that it calls attention to the dangers associated with their possession of authority, and indicates the pressing need for them to divest themselves of it. From this point of view, the problem is to encourage people to respond to scientists only through their reasoned arguments and demonstrable findings, and to desist from treating them as authorities to be passively relied upon.

A response of this kind comes very naturally and easily in modern democratic societies, where general modes of thinking are individualistic and anti-authoritarian. Milgram himself responds to his own results in this kind of way, quoting the words of the social philosopher Harold J. Laski: 'Our business, if we desire to live a life, not utterly devoid of meaning and significance, is to accept nothing which contradicts our basic experience merely because it comes to us from tradition or convention or authority.'[3]

And these are words, of course, which express ideals very firmly established in the thinking of scientists themselves and in the literature of the philosophy of science. The standard idea of scientific knowledge is, after all, that it is the result of a direct encounter between a normal individual and nature. A reliable observation is one which any individual with normal powers of perception might make. An acceptable law or theory is one which any individual with normal powers of reasoning might justify. Ideally, as it is generally perceived, the whole of science should rest upon such individual acts of perception and justification. And it should so rest because science is then based upon reason and experience alone, and can dispense with authority and dogma and similar evils. Science is then a form of knowledge to which people can be enlightened, not a form of knowledge which has to be pressed upon them or which gains acceptance out of credulity. It is a form of knowledge which can be accepted by anyone in a society of equal individuals, out of deference to nothing more than his or her own powers of perception and inference.

This is the *rationalist* vision of science, according to which each of us has access to knowledge by virtue of our own awareness of nature and our own reasoning powers, and each of us independently can act rationally and responsibly on the basis of what we know. It is an ideal vision of science and scientific knowledge which fits closely with the ideal of a democratic individualistic society. And it is a noble and profound vision; there can be no doubt about that. One has only to look back to Milgram's work to see what it seeks to avoid, what dangers lie in any alternative to it. Or one can look back instead to pre-scientific systems of knowledge, explicitly dogmatic and authoritarian, and to the forms of society which sustained them. Much of the heroic history of science (the kind which highlights Galileo's struggle against the Catholic Church and similar episodes) represents it precisely as an anti-authoritarian movement, liberating people from stifling pressures not just upon their actions but upon their understandings. For anyone with this view of science and its history the very last thing any scientist should expect is deference due purely to a white coat, or to a title before or letters after the name. Such an expectation would represent a failing, even an incipient corruption, in the scientific enterprise.

The problem is, of course, that it is very difficult to relate this ideal of science to what actually exists. We have a vast system of scientific education wherein, it seems, deference to the authority of the teacher is essential, as is a willingness to set aside one's own individual thoughts, ideas, and criticisms. And we have a very high level of intellectual division of labour within science, which again seems to require that one person defer to the perceptions, the reasoning and the recommendations of another purely on the basis of his standing, the letters after the name perhaps, and the cognitive authority they imply. I have indeed already conveyed in earlier chapters the way in which, in my view, the institutional structure of science is built upon relationships of authority. I have tried to show that research would not be a viable activity without such relationships, that if nothing else the whole system would collapse under an overload of information. But even supposing that this is wrong, that scientific research, contrary to appearances, does manage to proceed in some sort of conformity to the rationalist view, there remains the problem of how science interacts with other institutions and with the general public.

Consider the overall distribution of scientists and scientific specialists in our society at the present time and the range of work that they do. Consider them acting as consultants to industry and government; their role in medical diagnosis and medical treatment; their appearances in courts of law as forensic scientists and as other kinds of expert witness; their involvement in agricultural and environmental projects; their military relevance. Now drain away all the authority and trust implicit in all these activities, and ask whether they could possibly continue. The answer must surely be, no. Specialized expertise is coupled into its various, multifarious social uses precisely by means of relations of authority, and it is difficult to imagine what alternative means might be employed.

We do not habitually accept that trust and authority are a necessary feature of the distribution of knowledge in society because we habitually work with a very individualistic image of knowledge, one which takes no account of the intellectual division of labour in modern societies, or even of the more basic point that knowledge is always public property, always the possession of a society rather than a private individual. With an isolated individual as the focus of attention, knowledge and action based upon it, can be very simply related to each other. If an individual knows Euclid's geometry up to the twentieth theorem we can straightforwardly say that he is in a position to prove the twenty-first theorem: he knows all it is necessary to know. But imagine that this knowledge is spread over the members of a society, some known by some individuals, some by others. We cannot say of this society that it knows enough to prove the twenty-first theorem. To think of the society as an individual writ large in this way would be quite misconceived. Suppose that the different individuals, with the different necessary bits of knowledge, did not know each other, or how to find each other. Or suppose they did not trust each other, or know how to check on each other's trustworthiness. In both cases the twenty-first theorem would remain unproven. The technical knowledge would have been present in the society, but not the necessary internal ordering – the necessary social relationships – for the proof to be executed. Individuals would have known enough mathematics, but not known enough about themselves.

The knowledge, that is the particular competences and the individual beliefs, of the members of a society cannot be simply added up to define 'what the society knows'. There may, after all, be no way at all in actuality of ever adding up all the individual contributions. We should discard this way of thinking and look instead at how knowledgeable individuals are actually related to each other in the society, and how they interact with each other. What can be done with the various competences and beliefs distributed across society depends upon how the individual bearers of those competences and beliefs are related to each other, and how they interact. Knowledge in society is not simply so many ideas, concepts and procedures. It is systematically organized ideas, concepts and procedures, carried by systematically organized people. And the possibilities of the knowledge at any given time depend upon the organization of the people who are its carriers and its embodiment. This is the case whether one thinks of intellectual possibilities like proving a theorem or validating a conjecture, or of practical possibilities like building a nuclear power station.

Organization is built out of social relationships, and these social relationships involve trust and the acceptance of authority. It is precisely social relationships based upon trust and authority which allow specialized knowledge to flow through society to the points where it is used, and allow knowledge and competence from many origins to be combined at such points. As knowledge concentrates in specific sites in the social order, access to it is maintained through relations of trust and authority. What can no longer be called in by the individual from his own personal memory is called in from a trusted knowledge source. Trust and authority are the wires of a great system of communication which makes the specialized knowledge of society widely credible and widely usable. Knowledgeable individuals interact across these wires. The immediate knowledge-based capabilities of society are as much the product of this wiring as of the specific items of technical knowledge which this wiring connects.

Scientists must possess and use authority to serve as specialists in a society with high division of intellectual labour. And they do indeed possess it and use it. It is interesting to look back to the period when scientists first acquired this authority, and see what its acquisition involved. We must look back to the nineteenth century, to the time when science was establishing itself as a professional activity, and the term 'scientist' was coming into increasing use to describe its practitioners. At the beginning of this period cognitive authority lay predominantly with the Church; at the end it had swung to the sciences, at least in so far as knowledge of the empirical realm was concerned. The scientists had offered a secular, impersonal view of nature, had defended it against the criticisms of theologians and clergymen, and had successfully hit back in their turn at the claims and assumptions of the theologians' world view. There was for a period evident conflict and tension between scientists and clerics, each defending their own versions of the cosmos and calling the alternative into question. This was well recognized at the time. In England, for example, the second half of the nineteenth century saw something of a rash of books on 'The Relations between Religion and Science', 'A History of the Warfare of Science with Theology in Christendon', 'History of the Conflict between Religion and Science', and similar themes. Two versions of the cosmos had, apparently, come into collision, and it was the older sacred version which had emerged the worse from the engagement.

Why then was there this clash; and why did science emerge with most of the spoils? It might perhaps be argued that the new secular knowledge and understanding was simply superior, and attracted reasonable people away from alternative religious ways of interpreting reality. Certainly it can be argued, as I have already, that the culture of science and its thoroughly secular impersonal account of physical nature was extremely congenial to the rising classes of the new industrial society, and that their support played a crucial part in determining the outcome of events. Either of these ways of looking at the problem suggests that in the inevitable clash of cosmologies, of secular and sacerdotal world views, it was bound to be science which would emerge in the ascendant, and scientists upon whom cognitive authority would alight.

But there is more to the matter than this. Consider first some points which do not fit neatly with the above. First, when scientists began to organize themselves early in the nineteenth century, and develop many of the characteristic ways of thinking about nature which we now recognize as scientific, they did not see any clash of world views at all. Many of them actually saw what they did as supportive of Christian theology: they sometimes called their work 'natural theology', and thought that it served to show the operation of God's providence in the natural world. Secondly, even later in the century when many scientists did become explicitly critical of the pretensions of theologians and the relevance of their texts and doctrines, these same scientists remained, in most cases, themselves Christians: only the most militant and radical found themselves advocating agnosticism or atheism. Finally, at the turn of the century, when the engagement came to an end and scientists and clergymen largely abandoned their quarrel, it is hard to see precisely what ideas had triumphed, or what doctrines had been abandoned. The Christian religion plodded on; Anglicans accepted the same number of articles as before. Science continued its expansion; many of its members accepted the articles. The differences in cosmology had not actually been settled; no peace treaty had been drawn up. Everybody simply seemed to have lost interest in the old controversies.

All this has led the historian Frank Turner to suggest that the clash should not primarily be seen as a clash of ideas at all, but rather as a clash of professions.[4] When scientists became sufficiently organized, and aware of themselves as an occupation with occupational interests, they naturally sought social standing, patronage, financial support, and jobs. And a major obstacle to their acquisition of these things was that a good part of them were already in the possession of the clergy, the correlate of their cognitive authority in society. Inevitably, scientists and clergy found themselves in competition with each other. Scientists came into conflict with the interests of clergy in the course of their own expansion. And the clergy, recognizing the threat, found themselves opposing the pretensions of the new scientific disciplines, so that as T. H. Huxley chose to put it:[5] 'Extinguished theologians lie about the cradle of every science as the strangled snakes beside that of Hercules.' In a nutshell, the scientists recognized the need for cognitive authority, and actively set out to obtain it at the expense of the clergy. The clergy met the threat as best they could, but in the end lost out.

Looked at in these terms, there was no inevitable clash between alternative cosmologies. It was only when the expanding scientific profession found itself hemmed in by the older profession, the Church, and inconvenienced by its controls and constraints, that the picture of incompatible world views emerged, as part of the professional propaganda. The picture was used to criticize established religion and its public spokesmen and representatives, not to call into question private beliefs and speculations. It was used as part of the attempt to establish science as the legitimate repository of empirical expertise in society. Once the attempt was successful, and scientists came to possess what they regarded as a sufficient amount of authority, standing and support, the clash of the opposing cosmologies became less interesting and faded into the background. Clergymen for their part bowed to the inevitable and came to accept a reduced but still significant social role, taking care not to cross swords with the new breed of scientific experts on technical and empirical matters.

Turner documents a number of the strategies employed by the more militant professionalizing natural scientists in attacking clerical authority and seeking to enhance their own. One was to challenge all attempts by religious authorities to claim the capability of influencing or controlling empirical events. It is important to remember that claims of this kind were still being made in the nineteenth century, even at the highest levels of the most important denominations. Clerics often interpreted bad harvests, natural disasters and outbreaks of disease as divine retribution for sin and human wickedness, and suggested that observance of the moral teachings of the Church might keep them at bay. Failing this, people should look to the power of prayer, directed by their clergy, for the alleviation of their misfortunes and the improvement of their lot. God was to be thought of as managing His universe on interventionist principles, not those of *laissez-faire*: on receiving a memorandum or a petition from lower management, a bishop say or perhaps a cardinal, He might choose to act in the world to relieve an affliction or avert an incipient disaster. Turner notes how in the 1850s and 1860s prayers were formulated, at the highest levels of the religious hierarchy, to seek relief from the cholera, from an excessive rainfall at harvest time, and from cattle plague. And he documents how in 1871, when prayers for the Prince of Wales, lying stricken with typhoid, were followed by the rapid return of the Prince from death's door, the recovery was hailed as testimony to the power of prayer and turned into a triumph for the clerical interest.[6] This was just the kind of thing which the professional scientists found intolerable, since it bolstered the authority of the rival profession in empirical matters, and served to direct the attention of the public from the technical expedients which could offer them real hope of relief from their afflictions. Eminent natural scientists kept up a persistent challenge to claims for the efficacy of prayer; they even went so far as to propose, and in at least one case actually to carry out, experiments and statistical exercises designed to test the claims. Needless to say, the scientists saw no evidence of prayers' being efficacious. But the clerical interest found the notion of a statistical analysis of God's interventions in the world somehow inappropriate.

As well as attacking particular claims and assertions made by priests and clergy, spokesmen for science developed a conception of the nature of knowledge precisely designed to legitimate science and to call religious doctrine into question. This conception, which is now known as 'scientific naturalism', insisted that knowledge should be entirely based upon observation and experiment; that it should be devoid of all elements of dogma and metaphysics, and hence devoid of all purely theoretical terms and entities; that it should appeal wholly to reason and not at all to faith; and that its status should be that of a predictive instrument not a true revelation of the nature of the cosmos. Many outstanding scientists, of whom T. H. Huxley, John Tyndall, W. K. Clifford, Francis Galton and Joseph Hooker are perhaps the best known, used this conception of knowledge to extol science and devalue religious doctrine. Clerics made explicit appeals to faith and authority, and presumed to defend both metaphysical notions and claims to absolute truth. All of these things were grounds for criticism according to the doctrine of scientific naturalism, and all were exposed and attacked in the polemics of the natural scientists. It should be added that science itself was not systematically and unflinchingly examined in the light of the anti-authoritarian and anti-theoretical values of scientific naturalism: the main use of the doctrine was as a weapon against an external opponent. An anti-authoritarian ideology was being used as a weapon in a fight *for* authority: it was a good weapon because at the start of the fight it was the opponent who relied upon authority.

Finally, and perhaps most importantly, the new professional scientists sought after positions which would provide them with authority. In particular, they sought to penetrate into the educational system, into the schools and the universities, again at the expense of the clergy. Positions as teachers and as academics gave standing to themselves and their subject, permitted the reproduction of their profession, and furthered the aim of expansion into all the areas of society where technical expertise might be employed. But these positions had to be gained against the opposition of the religious groupings currently in control of the educational system: this involved calling for removal of theological tests in universities and informal requirements in the public schools, opposing denominational control of the school boards after the Education Act of 1870, and demanding that the science taught be science as defined by professional scientists'.[7] Francis Galton hoped that by usurping the position of the clergy in educational institutions, and by gaining responsible positions in the employment of government, it would be possible to establish 'a sort of scientific priesthood throughout the kingdom, whose high duties would have reference to the health and well-being of the nation in its broadest sense, and whose emoluments would be made commensurate with the importance and variety of their functions'.[8]

Overall, then, what Turner's work suggests, and indeed what a lot of related historical studies of nineteenth-century science suggest, is that cognitive authority did not simply fall into the lap of the emerging scientific profession. The scientists actually went out and fought for it. They worked to establish themselves as 'a sort of scientific priesthood', to situate themselves high in the social hierarchy – no doubt partly with a mind to the associated 'emoluments', but partly also because they recognized that such a position was essential if their views were to have credibility and their advice was to carry weight. The role of priest, a role which they attacked and criticized, was also a role which they emulated.

All this adds substance to what was argued earlier. Authority cannot be thought of as something which became attached to science by accident, or by outside agency, and which now poses a threat to the normal operation of scientists in society. Scientists have not imperceptibly slipped away from their proper role of convincing people by reason, into the easier but less reputable task of instructing them from positions of authority. From the very start, scientists built their position in society upon authority.

No doubt they saw no alternative. They were living in an intensely hierarchical society which simply worked in that way, and if they had rejected any standing as authorities others would have filled the gaps. But it is difficult to imagine that things could have been fundamentally different whatever kind of society the scientists were seeking to establish themselves within. The new scientific profession was increasingly a profession of specialists. It offered its services to society explicitly in these terms, as a repository of empirical and technical expertise. But expertise and cognitive authority are different sides of the same coin. It is difficult to see how there might be the one without the other.

If nobody is ever to accept the cognitive authority of anyone else, then everybody must have the same knowledge. On the other hand, if we are to reap the advantages of specialization, of an intellectual division of labour, then the direct testing and checking of what another person says must be replaced by a relationship involving trust and the acceptance of that person's authority, even if only in a restricted and carefully defined field. In a specialized society, where ideally everyone can benefit from knowledge that need only be in the possession of a few, the gain in efficiency is precisely the product of cognitive authority standing in for direct knowledge. A specialized society just is one with an elaborate and sophisticated system of cognitive authority.

The true dilemma raised by the existence of authority and its evident potency should now be clear. On the one hand it seems to be a key factor in maintaining our present distribution of knowledge and intellectual division of labour. And in a more general discussion, it could no doubt be shown to be essential in the maintenance of the division of physical labour and in the operation of the very highly ordered social institutions upon which we all rely. On the other hand, it represents a palpable danger – a danger which, when dramatized as in Milgram's work, tends to make us run a mile from any relationship involving authority and requiring obedience. One of the things which Milgram's work serves to do is to raise a question for everyone whose life is life in a specialized society: what can be the role of properly informed, fully responsible, individual judgements and actions in such a society? The answer must be, I suppose, that their role must be more restricted and less significant than might ideally be wished.

This, of course, is one of the criticisms advanced by the many people who dislike and deplore our current expert-dominated form of society and its tendency to move to higher and higher levels of specialization. And it is surely both a valid and an important criticism. But what are its positive implications? The difficulty in moving to a positive viewpoint, the difficulty which many critics refuse to face up to, is that being knowledgeable and being specialized are inseparable conditions. There are indeed a few people who are willing to accept a move back to ignorance as the price worth paying for the possibility of equality amongst individuals and an authentic way of living. This position is at least consistent, as also is the opposed view which accepts experts as part of the overall hierarchy of authority, and counts self-effacing obedience to authority as duty and hence as something which stands high amongst the virtues. But most people will seek an alternative to these two consistent positions. It is tempting to seek the best of both worlds, the advantages of specialization without its drawbacks, the benefits of authority relationships without their perils. This leads to the perception of authority as inherently *double-edged,* and to talk of its 'wise use' and of the importance of keeping it 'within the proper bounds'. Unfortunately, within our society there is no clear consensus upon what wisdom consists in here, or what proper bounds should be observed: we are indeed deeply divided on matters of this kind. And so long as these divisions exist our basic intuitive attitudes to authority itself are likely to remain divided also.

4

Expertise in Society

SCIENTISM

Science is now well established as the dominant form of cognitive authority in all modern societies: what counts as empirical knowledge in these societies is very close to being what scientists and associated experts allow so to count. But the authority of science does not have unlimited scope. It does not extend to the realm of morals. Many people still call it into question in the area of human behaviour and human choice. And there are any number of clearly empirical issues where the authority of science is actual but purely nominal. Think of next month's weather, or the problem of ageing and death. Or, more generally, consider that research is done in all the sciences on a large scale: this is a nice indication of the perceived inadequacy of existing knowledge and its inability to pronounce on all empirical outcomes. Massive ignorance must indeed be recognized as a normal state of affairs in all the natural sciences, reflecting the fact that we continue to live in an inordinately complex, incompletely known world.

At any given time in a given society, there tends to be a generally accepted view of the scope of scientific authority, an agreed account of what lies within its bounds and what does not. But this boundary only represents a kind of dynamic equilibrium: it stands where it does as the outcome of a contest between people seeking to push science outwards and others determined to press it back or to contain it. And because it is not a natural boundary, but merely the immediate result of the conflict of opposed pressures, it changes over time. Historically, the tendency has been for the scope of the authority of science to expand.

There are many current examples of natural scientists attempting to extend the legitimte scope of their knowledge and their techniques. The artificial intelligence (AI) movement, or more precisely a number of scientists within that movement, hold that human thinking and human intellectual capabilities can be satisfactorily described in the idiom of computer programming and symbolic logic, because in the relevant respects the brain is an information processing machine like a computer. A number of ethologists believe that what they have found out about animal behaviour can be extended to describe and explain human behaviour, and that there is no reason to exclude the human animal from conclusions meant to apply to animals generally.

Attempts of this kind to extend the authority of science beyond its currently accepted bounds are sometimes described as 'scientism', and the arguments used to support them as 'scientistic' arguments.[1] This terminology is in fact most commonly used by those who seek to keep scientific authority within bounds; for them it is a term of criticism or abuse. Scientism, for them, is the pretence of being scientific: a scientistic argument is one which involves an illegitimate appeal to science; a scientistic attitude is one which makes a fetish of science and wrongly treats it as the only possible form of understanding. But people differ on what is right or wrong, legitimate or illegitimate, scientific or scientistic: one group's science may be another group's scientism.

There is a very widespread and important method of criticizing allegedly scientistic arguments, which occurs so frequently that it is worth looking at in general terms. It is often said of such an argument that it is *metaphorical* or *analogical,* that it extends existing knowledge merely on the basis of analogy and that to accept it is to be taken in by an analogy. Consider, for example, the claim that the human brain is a form of computer, and that by the computer simulation of the intelligent activities of the brain it should be possible to learn how the brain itself is designed and programmed as an information processing system. Clearly the claim cannot be that the brain is literally a computer. Computers, in routine English usage, are physical objects with a keyboard and a visual display unit or tape output: brains are small flabby objects, generally found inside skulls. Nor do brains literally behave like computers: observing a computer and observing a brain are quite disparate experiences. Nor is it even possible to say that the human behaviour we attribute to the brain is observably that of a computer. By all the formal rules of correct description the brain is not a computer. To speak of it as a computer is to speak figuratively, to make an analogy. It is to say something which is formally incorrect, perhaps for the sake of convenient communication, as when we treat society in the image of the human body and refer to its 'health' or 'sickness' or to a 'rash' or muggings or an 'epidemic' of crime. It is reasonable enough, of course, to use analogies and figures of speech of this kind informally. But to be taken in by them to the extent of mistaking them for proper scientific descriptions is another matter. And this is just what occurs, so the critics say, when it is accepted that the brain is a computer, or even that it can properly be treated as a computer. To accept this is to mistake a suggestive analogy for a correct description.

The problem with this argument, however, is that 'mistaking' analogies for correct descriptions is actually a characteristic move in the development and extension of accepted scientific knowledge. Think, for example, of atomic theory, the theory which initially described gases as composed of small hard particles like billiard balls, flying through space, and described chemical combination as these same particles sticking to each other in fixed whole number combinations. This theory was a suggestive analogy. Initially it could have been criticized, much as the idea that the brain is a computer is now criticized. Indeed, a competent philosopher could still properly criticize atomic theory as a mere analogy: it would be nothing more than an academic exercise; nobody would actually wish to do it; but it could be done. But the science of chemistry developed precisely because people were disposed to make more of the analogy than perhaps 'strictly' they should have: in fact, the major thrust of scientific development came from people who accepted that atoms were *real,* and that the analogy was indeed much more than that. Chemical atomists went beyond what was 'strictly' legitimate in the way that they spoke of and thought about matter; but the result of this 'mistake' on their part was to change the rules of what is and what is not legitimate. The result is that today it is taken as a matter of course that we should refer to material substances in atomistic terms: molar solutions are part of the standard equipment of the chemistry laboratory; atomic weights are listed in the standard tables of facts.

The whole history of science is a story of the extension of empirical knowledge from one context to another by analogy. And at every point scientists have found themselves under criticism for accepting the 'mere' analogies involved. A major breakthrough in mechanics came when celestial matter was treated as analogous to terrestial matter, and the motions of the planets were held to follow the same basic laws as earthly objects. A similar threshold was crossed when organic chemical substances were treated as analogues of inorganic ones, and it was accepted that they were formed and held together in just the same way. Analogies between parts of the human body and mechanical systems have again and again become accepted after initial resistance: there was particularly strong opposition to overcome before the nervous system could be routinely treated in this way. Finally, the human brain has been arrived at, and this is the area where the problem of the legitimate scope of scientific knowledge still engenders real controversy. The brain (or, if preferred, the mind) is the last redoubt where those who cannot contemplate a complete surrender to hard-nosed scientific forms of analysis have dug themselves in, determined to resist all further 'scientistic' (or scientific) analogies and to prevent one vitally important area, that of human judgement and human decision, from becoming incorporated within the scope of scientific authority. Already, however, the wolves of neurophysiology and of AI howl hopefully at the door.

There is something intrinsically optimistic about scientific research. Scientists start with what they feel they know well and securely, and they proceed by analogy to related subjects – hoping that the analogy will prove fruitful. Without this optimism there could be no research. Accordingly, it is a malapropism to criticize all inferences which rely upon 'mere' analogy, and scorn them as 'scientism', since argument by analogy is crucial within science itself. Those scientists who insist that the brain is properly treated as a computer cannot be written off as unscientific propagandists after all. On the other hand, it is perhaps just as well that there are critics prowling the edges of science, looking out for scientism, ready to pour scorn on suspect analogies. Because analogies are analogies after all and can never be relied upon totally. Their standing must remain uncertain even when they have been shown to pay off time after time in practical use. And when such practical use is just beginning their standing must be very much more open to question. Those scientists likening the brain to a computer cannot sensibly be criticized for the research they do and the inferences they allow themselves as they do it. But that is not to say that these inferences have to be accepted as correct, and that the analogies they are based upon are valid. Perhaps the scientists themselves will eventually decide that the analogies are less fruitful than they had thought likely, and will abandon them. Who can say?

Argument by analogy is always less than completely trustworthy; and, worse, there is actually no fully satisfactory way of working out precisely how far to trust it. On the other hand, argument by analogy is necessary: there is no living without it. Inevitably, therefore people will employ it, and will accord it different levels of trust. Naturally, scientists will be liable to become enthusiasts for their own favoured analogies, and to put particularly favourable interpretations upon whatever evidence accumulates to support them: their trust in their analogies as their research proceeds will be liable to run ahead of that of other people. This is when they become particularly vulnerable to the charge that they are being 'scientistic' not 'scientific'. And this is when such a charge can perhaps serve a useful function, in challenging or even deflating what can sometimes be dangerous over-optimism or over-confidence. There can be no absolute standard to support a charge of this kind: one person's science can be another person's scientism, and there can be no way of finally settling the issue. But the collective argument about where the one ends and the other begins is most important for us, since its outcome effectively determines the legitimate scope of scientific authority at any particular time.

So far I have only considered attempts by recognized scientists to extend the scope of their authority. But 'scientistic' claims are also made by groups of outsiders, knocking, as it were, at the door of science, and seeking admittance. The core of accepted science is always surrounded by a penumbra of occupations claiming scientific status, but failing to secure it clearly and unequivocally. Such occupations range from the only slightly dubious, like psychology, to those which are liable to be dismissed as pseudo-sciences, like psychoanalysis and systems theory. Typically, one of the major concerns of practitioners in such fields is to gain acceptance as genuine scientists, along with all the associated benefits. In these fields concern with the fundamental characteristics of scientific knowledge and with the distinctive features of scientific method and scientific validation tends to be much more explicit and intense than it is, for example, in chemistry or physics. When one *knows* that one's field is genuinely scientific, there is much less call to ask what precisely that might imply.

The demands of outsiders for scientific status can often be intensely embarrassing for those firmly esconced in mainstream scientific disciplines. The standard argument of outsiders is that their knowledge or their method of working is entirely comparable to that accepted in science, and thus that they themselves are genuine scientists. And such an argument can prove difficult to dismiss, yet disastrous to accede to.

One such argument is that put forward by the creationist movement in the United States. This movement is made up of people who adhere to the Biblical account of the creation, and to the view of the origin of species and the origin of men set out in that account. They are, of course, aware that another account of the origin of species is widely accepted in their society, that of modern evolutionary biology, still sometimes known as Darwinism. The creationist view is incompatible with modern Darwinism: what the one attributes to genetic variation and natural selection the other attributes to God. But the spokesmen of the creationist movement are extremely reasonable people. They accept that creationism is but a theory which could be wrong, and that the knowledge of creationist biology is fallible and might eventually be found wanting. But, they say, is not modern Darwinism also a theory, and is it not the case that many of the findings and observations associated with it might also eventually be found wanting? Science after all is provisional, fallible knowledge, not absolute truth. So although, undoubtedly, modern evolutionary biologists deserve respect, and serious consideration for their arguments, they surely should not be allowed to stifle every other point of view, and impose their own like some monolithic totalitarian doctrine. Surely, the creationist viewpoint should be taken seriously also, and its proponents respected.

The argument is sweet reasonableness itself, is it not? It appeals to authentic scientific ideals, and to the spirit of tolerance and fair play which is so important in modern democracies. But there is a sting in the tail. What seems wellnigh impossible to reject as an abstract argument becomes impossible to accept as a practical proposition. Both modern biology and creationist biology are fallible theories, so the creationists assert, and both have a degree of support. Let them in that case coexist in a spirit of tolerance and equality, with both points of view presented in the schools and universities so that people generally are in a position to make a rational decision between them on their own account. Let biological education proceed according to a principle of equal time for both.

Think now of what this fair-minded suggestion actually means in practice. Think of the thousands of teachers who could face dismissal, the millions of dollars of research funds uncoupled, the unclear basis for future scientific advice and consultancy, the vast mutation in the general scientific consciousness of the nation. With practical issues of this magnitude at stake, abstract arguments take on a different significance. Needless to say, if the principle of equal time is ever extensively employed in biological education in the United States, it will be a consequence not of the power of argument but of a shift of political power of such a dramatic nature that it is difficult to believe that it could occur.

Creationist biology claims scientific status on the grounds that its knowledge is comparable to that of accepted scientific fields. But it is not even necessary to possess knowledge in order to lay claim to such a status. Many influential accounts of science represent it simply as a *method*, a way of obtaining and evaluating knowledge, not as knowledge itself. Parapsychologists claim to be scientists precisely on the basis that their methods and evaluations are scientific, even if, up until the present time, they have not actually yielded any knowledge, in the shape of reliable positive findings. A parapsychologist may be an ignorant scientist but he is nonetheless, so he believes, a genuine scientist.

Parapsychology is the field which seeks to study so-called paranormal phenomena, phenomena which according to the mainstream of accepted scientific knowledge do not exist. Among the best known of such supposed parapsychological phenomena are those involving extra-sensory perception, ESP, which may be manifested as an apparent ability to see distant places well out of eyeshot, or to become aware of what another person is thinking. Such apparent abilities have been systematically studied by parapsychologists and although no clear cut results seem so far to have emerged it does seem that the best of this work meets the highest technical and methodological standards: if there is indeed a 'scientific method' then this work comes as close as any to exemplifying it.

Parapsychology does indeed seem to be making some headway in its attempt to become accepted as a science. Several American universities have supported the subject, and there is now an established chair of parapsychology in a British university. Perhaps undergraduate courses in the subject will eventually become available; or even special government grants to develop the technologies which it will eventually open up. Nonetheless the subject remains something of an embarrassment for the mainstream of natural science. On the one hand, the argument that good method and sound inference is the key to science is hard to reject: scientists are scarcely likely to reply, in public debate, 'no, science is received dogma and inherited techniques'. On the other hand, to associate scientific authority with scientific method allows that authority to be claimed by practically anyone for practically anything. Parapsychology becomes the thin end of the wedge for astrology (currently doing surprisingly well), phrenology (still surviving), UFOlogy (thriving), futurology (a growth industry), and the scientific study of (possible) fairies at the bottom of the garden. I am not saying that a wide and thin spread of scientific authority would always and invariably be a bad thing, but I am saying that scientists in the established mainstream are well placed to see its disadvantages. Attempts to extend scientific authority of this kind, attempts made by outsiders, invariably stimulate formidable opposition from the mainstream of established science.

Finally, it is important to recognize that many experts or putative experts do not actually ever bid for the approval of the scientific establishment. They simply take up the trappings of science, its symbols and rituals, and thereby seek to clothe themselves in scientific authority. The advertisement where the man in a white coat recommends this bottle of pills, or that pack of tablets, is a convenient symbol of this simple form of scientistic activity. The whole of society is, of course, shot through with it, since science is our accepted repository of cognitive authority and the temptation to borrow it is ever present. But opinions differ as to its importance. Some people think that it is a trivial phenomenon, widespread perhaps but small beer nonetheless. Others think that an undue and far-reaching influence can be exerted simply by the manipulation of the empty symbols of science. They believe that the white coat does sell the pills, and that it sells a lot else besides. They paint a depressing picture of politicians drooling over computer models, however fatuous, and of managers, bureaucrats and other decision makers, unduly influenced by futile quantification, laughable applications of statistics, fancy language, and so forth.

There is an important issue here. How far is our policy making and decision taking, and indeed our everyday life generally, based upon reliable expert knowledge, and how far upon dubiously grounded advice sustained by the puff and pretention of its presenters? How far are experts used and credited in a way which reflects 'what they really know', and how far in response to their style, to their manner of performance of the expert role? If one thinks of highly technical expertise, like that of the physicist and the engineer, and if one thinks of the demand for that expertise arising only in situations where it offers immediate and clearly visible pay-off, then it seems unlikely that spurious expertise could sustain itself for long. Demand for it would surely die away as people found its prescriptions leading to disaster. The ignorance and incompetence of the spurious expert would be bound in the long term to show through his thin scientistic veneer. No doubt the odd hundred million here and there does regularly go down the drain due to heed being given to such spurious technical expertise; but it is unlikely that it counts for much in comparison to the thousands of millions which policy makers likewise send down the plug-hole through the mistakes of fully competent experts, pay-offs to mollify and appease vested interests, and good, straightforward, old-fashioned corruption.

It is probably a mistake, however, to think of all expertise after the model of the physicist or the engineer. A fair amount of 'expert' knowledge is not put to technical-instrumental use at all, or if it is, not in a way which allows easy judgement of its efficacy or predictive power. Whereas natural scientific experts generally provide technical know-how, some experts find themselves merely called upon, in practice, to provide justifications and legitimations for various kinds of actions. Anxious parents read the latest works on child development and child rearing. It is no good their waiting for well-validated knowledge; the children must be reared. Nor will the latest works venture to predict how the children, their particular children, are going to turn out. But by following authoritative advice the parents do at least have the justification and the reassurance that they did the best they could, whatever the outcome. Courts of law listen to the evidence of psychiatrists and forensic scientists. However much they might like to, they cannot await the development of these fields before they decide one way or another in given cases: sane or insane, guilty or not guilty. But by hearing expert opinion, such as it is, whatever on earth it is, and then pronouncing their verdicts, the courts may be said to proceed in the light of the most authoritative available opinions: the hearing legitimates the judging.

In these kinds of context the expert role must be performed whatever the state of knowledge. The great institutions must continue to grind on. Defendants must be tried; the mentally ill identified, confined and treated; the young educated and trained; the bank rate set; whatever it is that experts 'really know'.[2] But given the kind of society we are, with our respect for science and expertise, there is nonetheless a demand for experts in all these contexts. And the cynical, but nonetheless probably the correct, hypothesis must be that where the demand exists 'experts' will appear, conjured into existence by the need for their presence, without, in this respect, what they 'really know' being salient. Moreover, the natural mode of this kind of expertise will tend to be that of scientism. The whole demand for it is after all a demand for authoritative pronouncements; only authoritative pronouncements serve to reassure or to legitimate. And to take on the appearance of authority in this way implies taking on the appearance of science.

There is now a considerable range of expert occupations which are, as it were, weak on know-how, strong on image. They are well established, with important roles in our institutional structure, a significant audience amongst the general public (as can be seen from the bestseller lists), and fine career and salary prospects. It is common to make adverse comparisons between the knowledge of the natural sciences and the knowledge of these occupations. But the comparison should not blind us to the manifest practical success of the latter groups. They may not have won universal respect for their stock of knowledge, but they and their practices have nonetheless thoroughly permeated through the texture of society; like mould through a cheese.

TECHNOCRACY

So far, in discussing our relationship to technical, scientific experts, I have argued as if the reliability and efficacy of their knowledge is all we need to think about. I have stressed the immense difficulties involved in fixing the scope of a body of expert knowledge, the range of situations in which it will prove reliable and effective. But I have tended to assume that where it is reliable the experts bearing it should be trusted and invested with authority – as though reliability were the only relevant issue here. This, indeed, is our usual way of thinking about specialists, and their usual way of thinking about each other. We ask what subjects a specialist is qualified to talk about, and we give his opinions a particular respect when they concern those subjects. But there is another way of thinking about the matter, one which was very briefly touched upon at the end of the previous chapter: we can look at the overall distribution of expertise and at the consequences, for good or ill, of that distribution. If we adopt this alternative way of thinking and ask whether or not experts should possess authority, we are not then primarily asking whether they can be relied upon, but whether or not it is desirable to be so reliant upon them.

This is, I think, a very important and interesting way of thinking about expertise and I want to try to convey some sense of what it involves. To do this I shall refer to some ideas developed many years ago now by the German writer Jürgen Habermas. However, although Habermas must take all due credit for the ideas I am going to talk about he must not be blamed for the precise way in which they are presented: I want to give as simple an illustration as I can of one or two important themes, not a careful account of Habermas's thought itself, which is complex and difficult.[3]

Consider first then the way that specialists and experts have proliferated in the developed countries over the last century or so. Scientists and technologists have burgeoned forth in ever-increasing numbers, and so too have professional administrators, lawyers, economists, and the like. These people have mainly been absorbed into the great bureaucratic structures of society, and in particular into the state bureaucracies and other forms of state employment. One of their main functions has been to advise policy makers and decision takers, often elected politicians, so that the most effective means can be adopted for the furtherance of political ends. Expert technical know-how has been put at the service of whatever political faction has been elected to power, and required to serve the political objectives of that faction. Habermas calls a society of this kind, wherein specialists are 'on tap but not on top', a 'decisionistic' one; it probably represents how most of us imagine our own society to be at the present time.

Now, to describe the distribution of power and the different forms of political experience in such a 'decisionistic' society is much the same thing as to describe the distribution of knowledge. At the top of the system is a political elite or leadership, making decisions to further its ends or interests. It has access to a vast array of specialized knowledge and expertise, set out below it precisely to facilitate and inform its decisions. The elite does not itself need to possess highly specialized skills and competences, but must merely be sufficiently knowledgeable to evaluate the great range of technical information offered to it. Its access to that great range of information puts it into a unique position. It is at once the most informed and the most powerful sector of society. And in its experience politics and life generally are strongly and meaningfully related, indeed they are close to being one thing.

Immediately below the political elite is a layer of experts and specialized administrators whose technical knowledge and skills gives them a privileged position in society, and a real although subservient involvement in decision making. These people, who know their own fields in unparalleled depth, but who as individuals are limited in their knowledge to just one narrow purely technical or purely administrative context, enjoy a corresponding degree of power and influence. In their experience, politics and their specialized professional identities are meaningfully related – but not politics and their life generally (unless they actually choose, as many do, to make the whole of their life revolve around their specialized identity). In general, these people are well enough rewarded, and possessed of sufficient influence, to be firmly committed to the existing structure of society. But the price they pay for their influence is that they must pass their technical advice upwards to separate decision makers, foregoing any explicit rights in policy making and helping to starve the remainder of society of information. They are expected to conform to norms of anonymity and confidentiality.

Finally, there is the third section of society, the great mass of the general public. Since in modern societies political decision making is a technical business demanding specialized knowledge and competence, and since the public both lack this knowledge and tend to be denied access to it, they are effectively cut off from any real involvement in ongoing political activity. Because they are ignorant, they are powerless. Habermas refers to them as 'depoliticized'. Their participation in the political process tends to be restricted to the periods before general elections, when on the basis of restricted and distorted information, filtered by the media, degraded, trivialized and biased by advertising agencies and professional communicators, they choose between competing political elites. Not surprizingly, therefore, many among the main body of the population perceive a sharp disjunction between politics and life generally, and become deeply alienated from their political institutions: occasionally there is active hostility to them, more often complete passive indifference.

In the 'decisionistic' society then, three levels of knowledge are three levels of power: the power structure and the distribution of technical knowledge are the same. It is their access to technical expertise which gives political elites their power. They are able to use that expertise both to arrive at decisions, and, perhaps even more importantly, to legitimate and justify them once made. The two activities are very different. When taking a decision, technical matters are weighed against each other, the pros against the cons. But when justifying a decision the pros are stressed and the cons are minimized, or even ignored altogether. Politicians and decision makers are in a position to suppress the advice which counts against what they are doing and to emphasize that which supports it. An individual specialist adviser, working with integrity in complete good faith, may find his work actually functioning solely and simply to legitimate and justify current government policies, inspiring ministerial speeches when it serves to support policy, filling the waste-paper basket when it does not.

Clearly, effective opposition to government policies legitimated in this way is only possible for opponents with access to their own experts; and they will in turn select advice which questions policy and overlook that which supports it. This means that controversy becomes restricted to elites, cliques and lobbies, operating, as it were, at the apex of society, and that armies of experts produce technical reports, surveys, computer simulations, calculations and so forth, to be employed almost entirely to legitimate the cases of the different sides. The exchanges between politicians in the context of the Common Market offer many examples of this kind of thing. Similarly, when a government seeks to purchase a new advanced weapon, and opposition criticizes the decision, it will almost invariably be found that the weapon will cost more, and function less effectively, on the expert assessments cited by the opposition than it will according to those cited by government.

Habermas finds the increased use of technical expertise for the task of legitimation an alarming development. He is not primarily worried by the threat of bias in expertise, or even by its misuse. Nearer to the heart of his concern is anxiety that technical issues have overwhelmed and swamped all others, that in devoting attention to increasingly complex technical debates we are diverted from more significant and fundamental issues and even start to lose our capacity to deal with them. Think of the way in which the introduction of a new weapon is now considered and appraised at the 'higher levels' of society. Nearly all the appraisal will deal with the weapon as a *means*, with the question, for example, of how cost effective it is in killing people, and how long it is likely to retain its capacity to kill people. Debate about ends, about whether it is right to possess the capability to kill, or to kill the particular people involved, civilians say rather than military personnel, will be far less extensive, to say the least, and less visible. Moreover, amongst the specialists and decision takers, technical debate will be couched in the language of experts, and its outcome is likely to be presented in quantitative form, whether it consists in a statement of weapon performance or a statement of its cost. This may imbue the technical debate with an aura of authority and reputability which debate about ends cannot match. We lack recognized experts in ethics, morals and general human decency, and there is no special language in which to discuss these subjects. Moral argument is thus far too easily perceived as pointless and futile, since 'it is all just a matter of opinion anyway'. And people may turn from it accordingly, and look to technical problems instead, where the issues 'can at least be rationally formulated and resolved'.

The proliferation of technical experts, therefore, may cause people to lose confidence in themselves, in their own intuitions and common sense, in their own informal ways of talking together and developing agreed notions of right and wrong. And with this loss of confidence they are obliged to fall back upon specialist opinions, the grounds of which are beyond their understanding, and which concern in any case only the instrumental aspects of political problems and not the far more important practical, moral aspects.

Habermas not only laments what he sees as this unfortunate aspect of our use of expertise, he foresees circumstances in which matters might become a good deal worse. Imagine that science and technology continue to become ever more elaborate and mathematical, and that society itself continues to become more complex, so that economic and administrative skills have to become more and more sophisticated. This would draw more and more technical experts into the service of the state and the great bureaucracies, and force decision making to rely more and more upon very elaborate technical analysis and technical advice. It could well be, in these circumstances, that the technical experts would move beyond the control of political leaders altogether, that they would cease to be 'on tap' and become instead very much 'on top'. Political leaders might find themselves no longer competent to evaluate technical advice, and forced accordingly simply to trust what their advisers told them. And as the political leaders became in effect ignorant, so they would become powerless: they would find themselves acting as front men for permanent experts and bureaucrats, carrying out public relations exercises for those at the head of administrative empires and industrial organizations. Habermas would call a society wherein this kind of development had occurred a *technocracy*.

In a technocratic society, control is in the hands of experts and administrators: affairs are ordered by some kind of scientific intelligentsia. The great mass of people remain depoliticized. Knowledge is distributed to define just two groups, the experts and the rest. And again the distribution of knowledge is the power structure: the experts have power and the rest do not. The top group of the 'decisionistic' society, the political elite, is no longer of major significance in a technocracy, and the selection of political representatives in general elections is hence of even less significance than before. Ordinary people are likely to feel even more alienated from political processes.

How though is it possible for political elites to lose their function? It is they who define political aims and objectives. A society cannot continue to operate without aims and objectives; who is to set them if not political leaders? The answer is that the experts and administrators themselves set the goals of a technocratic society, and in a particularly insidious way.

As the dominant element in a technocratic society, we may assume that technical experts and administrators are likely to be highly satisfied with it. Their own concern is likely to be with the stability of 'the system', with keeping everything ticking over smoothly, perhaps with a spot of economic growth and the occasional small adjustment and improvement here and there. Needless to say, this will not be the concern of everyone. People at the bottom of the system may well fancy more far-reaching changes. But the experts are well placed to legitimate and justify their own ends by making use of their standing as authorities on technical issues. How is this possible? How can standing on technical matters be exploited in pronouncing upon moral and evaluative matters? How can standing concerning means give authority concerning ends? The trick is to talk in terms of what is *feasible,* what is technically possible. Any clamour for profound and far-reaching change can thereby be dismissed as impractical, or ridiculed as utopian; and the limited, conservative ends of the technocrats can be made to seem in some way objectively necessary: there is not the money for anything else, or the technical means, or a way of proceeding which will not involve sacrifice or danger elsewhere. The more people become used to perceiving controversy purely in technical terms, the easier it is to get away with this kind of legitimation. To the extent that it is successful it transforms the large problems of how people should properly live together in society into small problems of management and maintenance. Society becomes perceived as a kind of smoothly operating machine needing regular servicing and occasional repair.

Needless to say, Habermas himself has no affection for technocracy. It is a society where the knowledgeable dominate the ignorant, where an unequal distribution of knowledge implies an unequal distribution of power. And above all else Habermas is a writer who abhors domination and seeks for an end to it. Any form of technocracy must be anathema to him, even a technocratic Brave New World where all needs are met and all individuals actually feel content.

How though might domination be eliminated from society and the tendency to technocracy reversed? Habermas's analysis implies that it is unequal access to knowledge which sustains an unequal distribution of power, and that if power is to be diffused knowledge must be diffused as well. But Habermas is at the same time appreciative of the benefits brought by modern science and technology and well aware that knowledge in those fields is the product of specialization. Thus he has to face the dilemma we encountered earlier in discussing specialization, and he faces it in a particularly acute form. How do we combine the benefits of specialized knowledge and expertise with life in a society where no sector is unduly dominant, no group unduly pressed upon by another? This is his problem.

It is a problem which neither Habermas nor anyone else has even begun to solve, even at the level of theory. It could be indeed that the problem is an insoluble one, although Habermas himself has always refused to countenance such a pessimistic view. His hope is that increased interaction and genuine dialogue between technical experts and the general public will effect a movement in the right direction, so that on the one hand technical decisions will become more informed by common sense knowledge and more sensitive to a wide range of interests, and on the other hand common sense will itself be enriched by incorporating something of the knowledge and the ways of thinking of the technical experts. But although this would indeed be a highly desirable development, offering untold benefits for any society wherein it occurred, it is hard to see how it could solve the fundamental problem. If inequalities in the distribution of knowledge, or in the distribution of access to knowledge, generate and sustain inequalities in the distribution of power, then equal access to knowledge at least must be necessary to produce the kind of society which Habermas seeks. And nobody has yet been able to show how this could be achieved whilst at the same time retaining our specialized bodies of knowledge and the expert practitioners associated with them.

ANOTHER VIEW

Habermas takes a large view when he writes on expertise. He looks at the whole system and tries to understand what it implies for our way of life generally. He thinks of knowledge not just in the traditional way, in terms of validity and efficacy, but in terms of the wider consequences of its existence and of the way it is distributed. In particular, he perceives that there must be a systematic connection between knowledge and power in society, and he attempts to understand what it involves. I think it is very important that this way of thinking about knowledge and expertise should be taken seriously, and encouraged to coexist with our more familiar piecemeal ways, but it is very uncommon in the English-speaking world. This is why I had to turn to a European writer to exemplify it: for some reason writers on the Continent are far more ready to take a large view than we are.

It has to be remembered, however, that any large view of knowledge and society must also, inevitably, be a speculative view. This is certainly true in the case of Habermas. His writings are important, but whether or not they are correct is quite another matter. Certainly, his premonitions of technocracy deserve to be treated with suspicion. For us to be dominated by technical experts there must be a high level of coherence, and hence community of interest, amongst the experts; they must act in some sense as a unified source of power or authority. But it is far from clear that experts are, or ever will be, bound together in this way.

It must be conceded to the 'technocracy' hypothesis that a large proportion of our technical specialists is found in the higher levels of the state apparatus, and that much of the remainder is strategically situated high up in the military and industrial bureaucracies. To some extent, also, it makes sense to treat the members of this body of experts as natural allies, united by many shared interests and shared ways of thinking. But as soon as we look at any important concrete problem of policy in our society what we invariably find is divided opinion and *divided expertise.* The role of nuclear weapons in defence is an obvious example. The choice, development and siting of nuclear power stations is another, closely related in the public mind. There is a vast range of such issues in the areas of health and safety. People are divided as to what to do about tobacco addiction, given its relationship to cancer of the lung, heart disease, emphysema and premature death generally. There is concern about lead in petrol, about the manifest dangers associated with so many widely used drugs, about the side effects of the pill, of X-rays, of water fluoridation: in all these cases people tend to weigh the benefits and the snags of a practice differently and to differ in consequence upon whether to ban or encourage it. Much the same is true of many projects which affect the environment, offering economic benefits perhaps at the risk of increased pollution or ecological imbalances: some people prize the benefits of conservation; others have no reservations about sacrificing them for economic gains. In all these cases technical expertise is involved upon both sides. And the involvement is such that different political objectives are sustained by different versions of the facts and different interpretations of their significance.

Obviously the list could be extended and sub-divided indefinitely. Scarcely anything done in a modern society goes unchallenged by some group or interest. And in pratically every case where the issue is significant and the challenge has any weight technical experts become involved, and not just as consultants and providers of expertise but in most cases actually as advocates and committed defenders or opponents of policy as well. This is something with which we are now all perfectly familiar, which has come to seem normal and natural. When experts appear on the television we are prepared for them to appear in pairs, with a chairman in between. We recognize the format and we await a confrontation. The format is precisely that used for party politicians. And just as we expect the Conservative to fend off his Labour opponent, so we expect the representative of British Nuclear Fuels to try to fend off the one from Greenpeace, or the representative of British Agrichemicals to deny the charges of the spokesman of Friends of the Earth. Much the same is true when experts make appearances in more specialized contexts. Courts of law expect to hear from at least two experts at a time, one for the prosecution, one for the defence. Public enquiries operate analogously.

One of the factors which sustains differences in the opinions of experts is their different forms of professional training and their different professional allegiances. Every expert profession maintains particular ways of perceiving and ordering phenomena and particular means of interpreting their significance. These ways and means are used to produce an account of any specific problem. But there is no guarantee that the different accounts emerging from different groups of professional experts in this way will all be compatible with each other: it is perfectly possible for them to be contradictory. And when they are contradictory we must expect technical controversy with both sides defending their characteristic methods and their accepted forms of interpretation, and calling into question those of the other side. It is important to recognize that most technical controversies have the form of a competition between two plausible *interpretations* of a situation. It is unusual, although very far from unknown, for them to involve explicit fraud and deception, or straightforward differences over data of a kind which might readily be checked. Thus, conflicting technical opinions are typically very difficult to decide between, and technical expert controversy has many of the features of theoretical controversy in science, briefly alluded to in chapter 2.

A recent case study has suggested that much of the early controversy about the effects of lead additives in petrol can be related to the different kinds of training of the technical experts involved.[4] Among the professional toxicologists who initially considered the question a 'threshold' theory of toxicity had long been paradigmatic. In their frame of thinking, the key thing was to keep lead levels well below the point where they saturated the excretory mechanisms of the body: above this threshold lead intake would exceed lead excretion and an accumulation of lead would result in the recognized symptoms of lead poisoning. Lead additives in petrol, however, were regarded as incapable of moving lead intake levels to anywhere near the threshold value, so the toxicologists, almost unanimously, were confident in regarding them as safe. In 1965, however, a geochemist explicitly criticized this work, and tried to set the problem in a new context. Lead levels in the environment of modern societies might indeed, it was agreed, be far below the threshold for acute lead poisoning, but they were well above 'natural' levels as indicated by geochemical techniques. It could scarcely be considered safe to allow the general population to wander around in an environment stuffed up with several times the natural level of a toxic substance. People might not be dropping dead everywhere, but they were still being subjected to 'chronic lead insult', which could well be having adverse if non-lethal effects. A vicious controversy followed the publication of these claims. And much of it was concerned not so much with the details of the two opposed interpretations as with the authority of their carriers. Toxicologists asserted their status as the most experienced and competent assessors of toxic risks, and the centrality of their concept of threshold to such assessments: opponents derided both claims. A complex debate about the effects of different levels of environmental lead was also very much a professional demarcation dispute.

Not only professional backgrounds and professional allegiances divide experts. Every technically skilled expert is also a private individual with a unique background and a unique range of private affiliations and commitments. Thus, for entirely personal reasons, individual natural scientists have at times actively sought to aid defence research and to contribute to work on nuclear weapons or biological warfare; whereas others, in the same way, have given technical aid and assistance to unilaterialist and pacifist movements.

It is hard to say how much positive importance should be given to the variety of personal commitments found amongst scientists and experts generally. But it does have a very real negative significance. Because of the diversity of these commitments, because scientists come from all kinds of backgrounds, make all kinds of affiliations, and possess all kinds of moral and political views, it is very difficult to organize them all for the furtherance of any important political end. Many attempts have been made in the past to organize scientists as a coherent political force and systematically direct their influence to the achievement of specific political objectives. The Association of Scientific Workers, the Society for Freedom in Science, the British Society for Social Responsibility in Science, Science for People are just some of the societies which have been founded in Britain for this kind of end. But no such society ever seems to recruit more than a proportion of practicing scientists, whilst at the same time engendering active opposition from a comparable number, so that its overall political effect is very small.

Like most other occupational groups, scientists tend to disagree on outside issues. They differ in their assessments of the merits of capitalism and socialism, the free market and state intervention, multilateralism and unilateralism, and practically every other major political issue. Only upon such immediately relevant political aims as increasing the pay of scientists, and the amount of support for their work, do they tend to be united. And these are the kinds of issue which unite any occupational group, coal-miners, or dairy farmers for example. It is curious that persistent attempts are nevertheless made to organize scientists on much larger issues, as though a wider-ranging moral and political uniformity did actually exist in the world of science.

Finally, and probably most important of all, experts are divided according to who employs them or finances them. Except perhaps for the significant proportion of academic scientists still financed at the longest of arms' lengths, the source of their support or employment inevitably places a systematic pressure on experts. The extent of the pressure varies. In some cases it may merely produce a degree of caution and reserve when a scientist bites the hand that feeds him. But in other cases, indeed probably in most cases, constraint is far stronger. It may well be taken for granted that a scientist in government or industry will never 'go public' on any matter, unless it is at the behest of his employers and in order to defend their interests. And employers are in a position to enforce this expectation, as several experts know to their cost: both in Britain and the United States, for example, experts in the nuclear industry who have made their technical reservations public knowledge have rapidly found themselves without employment. It takes exceptional individual courage to blow the whistle on a large organization in this way, even if one follows what is now the standard advice: 'first line up a good lawyer then line up a good job.'[5]

Given the degree of control that employers can exert over technical experts, it is not surprising to find that they are now frequently used much as lawyers and attorneys are, as little more than mouthpieces. 'It is virtually standard practice on the part of large technical organisations today to do "answer analysis" – that is, produce analyses in support of preselected positions – even during the internal decision-making process.'[6] In an organization which uses 'answer analysis', a technical expert may be directed to prepare a case in favour of some new project, to argue, for example, that it is safe, environmentally benign and profitable, by senior administrators who simply see a possible future need for such material. Or he may be directed to make the opposite case, or possibly to make both cases so that the organization can use either according to whether it subsequently decides for or against the project.

It might be thought that scientifically trained experts would react with revulsion to the demands of such a role, and would see the tasks of advocate and technical adviser as incompatible. But there seems to be no shortage within any organization of advocate experts happy to bolster its image, defend and justify its products, dig up dirt on its rivals and their products, and so forth. We are all now familiar with the clash of advocate experts of this kind. It is a direct reflection of our social structure and the conflicts between the institutions and organizations within it. Government expertise will be challenged by industrial expertise; specialists at the Ministry of Defence will not see eye to eye with those at the Ministry for the Environment; consultants for the Milk Marketing Board will tend to differ on technical matters from those employed by Unilever.

The advocate expert is now a commonplace, routinely accepted figure, so it must be presumed that the role has become accepted as a reputable one. Perhaps such experts justify what they do with the thought that 'the other side' has its experts too. Just as courts of law may get a fair picture from two opposed biased counsels, so a fair picture may emerge from two biased experts. Indeed may this not be the way that expertise *must* be deployed, given that we now live in such a highly differentiated, specialized and fragmented society?

There is probably something in this type of rationalization. No doubt, in our kind of society organizations must have technical experts prepared to point out their side of a case, to make sure that arguments and evidence in their favour are not overlooked, and that spurious attacks upon their products are exposed for what they are. Without this kind of technical resource organizations would be destroyed by their enemies or competitors. But it must nonetheless be hoped that the rationalization does not have too much success. There are some jobs in society, even essential jobs, which must continue to carry a bad smell with them. And that of the advocate expert is one such. We need the stink to remind us of how readily the job undermines integrity and invites corruption. Anyone in the role of an advocate expert, simply and solely because of the role he is in, must be treated with intense suspicion and distrust. There may be no limit to what interested experts will do on behalf of their employing organizations. One has only to think of their activities over many years on behalf of the tobacco industry, or of the long delaying action they so skilfully carried out for Chemie Grünenthal, the German manufacturers of thalidomide, in order to grasp this. On such an important matter it is necessary to be blunt. Quite literally, many experts of this kind are killers. As far as assessing what they have to say is concerned, historical evidence leads to the same conclusion as simple common sense: they should never be given the benefit of the doubt.

Let me return, however, to my main theme. I have been trying to convey a sense of the extensive divisions amongst experts, and of their significance. I have pointed to divisions arising from professional training and allegiance, divisions stemming from individual commitments, and divisions relating to employment and sources of support. And I have shown how all these divisions can lead to conflicts of opinion, to the extent indeed that almost every significant policy issue currently confronting us is enveloped in a cloud of technical controversy. The picture which emerges, it seems to me, is not one of dominance by technocrats, or even of a likely movement to such dominance. The technical experts are too fragmented, too strongly attached to the interests of other powerful sectors of society, and hence too prone to speak with more than once voice.

If there is to be a technocracy, if knowledge is to be the basis of the power of technical experts, then the knowledge must be uniform and the experts united. When one expert speaks against another then the general authority possessed by experts has no significant effect. The power to determine *which* expert is believed is the important form of power in such a society, not the power of experts themselves. There is also the further question of whether even near unanimity amongst experts makes what they say compelling. Certainly, there are cases where this has not been so. The general public seem to have consistently underestimated the dangers of nicotine addiction, and overestimated the dangers of cannabis use, in relation to widely accepted expert opinion. They have rejected expert findings about the effects of capital punishment, and of pornography. Even upon the apparently bland issue of the safety of water fluoridation a near consensus amongst experts has been rejected by many communities in the United States and in Europe, and many fluoridation programmes have been halted in consequence.

Given all this one is tempted to ask whether it was sensible to speak of the authority of science in the first place. It was, I think, indeed sensible. The authority of science quickly becomes manifest if one attempts to proceed without it in a modern society. But it is a limited form of authority, and insufficient as a basis for independent power. There are many points of similarity between the standing of scientists and technical experts in society today and that of priests and clerics a couple of centuries ago. In those days cognitive authority was vested in men of religion. Their specialized knowledge was knowledge of a moral order: it entitled them to specify the appropriate ends of human action, as well as the appropriate means. Did this mean then that society was dominated by its priests, that they could do as they would? The answer to this is negative. Society was dominated by religion, yes; but by its priests, no. There were, after all, different denominations of priests and different shades of priestly opinion within each denomination. And these denominations and opinions to some extent were reflections of different opinions in the wider society. The great weight of priestly authority did of course lie at the disposal of the dominant groups of society, but some priests of some denominations did take the side of the common people against established authority, clerical equivalents of Tribunes of the Plebs. And as the social order changed so too did the number and the relative significance of different denominations, with rising social classes being accompanied by their own favoured forms of doctrine and ritual. In what was a religious society all significant issues were expressed in a religious idiom, all conceptions of life in religious practice. Priests, for all their authority, had to sense and express opinion as well as to attempt to shape and form it. By analogy, we might say that modern society is dominated by science; but not by scientific experts. Expert assertions today must be expressed in a scientific/technical idiom; that is essential, just as centuries ago a religious idiom was essential. But it no more guarantees that a scientist will be believed today than it guaranteed that a priest would be believed long ago.

On the whole our society seems more like what Habermas calls a 'decisionistic' society than like a technocratic one. Technical experts remain 'on tap' and have made no attempt to convert their cognitive authority into political dominance. The suggestion that the distribution of knowledge in a society is also the distribution of power is, I believe, wrong – although the claim that the two distributions are related is surely correct. Knowledge is invariably easier of access to those at the top of society than it is to those at the bottom, and it does indeed, in so far as it serves expedient interests, predominantly serve those of the establishment and the dominant economic and political elites. It is true that in recent years experts have increasingly come to the assistance of grassroots pressure groups and other elements of the general public and local communities. And it is equally true that the grassroots have learned how to deploy technical expertise, and have exploited it with some success in furthering their interests and in embarrassing the establishment. But basically so long as the top of society remains greatly superior in resources it must continue to enjoy greater access to technical expertise and a very advantageous position in the playing-out of technical and thus political controversies.

Whether or not this situation is found satisfactory will tend to depend on one's general political perspective. One view of society has the top systematically exploiting the bottom, and sees the divisions of interest at the top of society as minimal and of no fundamental importance. From this perspective the concentration of expertise at the top permits the exploitation of the ignorant, depoliticized masses of society, and is thus highly undesirable. But there is another equally significant perspective which sees the elites at the top of society as representatives of factions within its main body. From this perspective the divisions at the top over policy, divisions reflected in controversies between technical experts, are real and deep: they are the natural consequence of all the diverse general interests which are represented at the top. This *pluralist* account of how society is ordered might encourage us to see the conflicts between technical experts as reflections of important social and political conflicts, and hence in a sense as a good thing. When many interests are represented at the top, so it is argued, no one interest may become overwhelmingly predominant. And if the consequent tensions, manoeuvering and jockeying for position at the top of society give rise to the use of interested, partisan expertise, and hence perhaps to some inevitable abuse of expert knowledge, how much the worse might it not be if there were no such tensions, if the top were indeed united against the bottom. This is the key necessary condition for totalitarianism.

Ultimately two different conceptions of human freedom are operative here. One sees it as a possibility which can be realized only with the advent of the ideally ordered society, with the demise of all hierarchy and all exploitation. The other sees it as a partial and precarious achievement which is available to individuals only so long as society remains fragmented, disordered and heterogenous, and which can be preserved only by ensuring that no monolithic organization is allowed to grow up as a source of future repression. Many moral and evaluative problems in our society require a choice to be made between these very different conceptions.

5

Thoughts on the Future

POSSIBILITIES

People in all societies speculate about the direction of social change and about the opportunities and the problems which are likely to confront them at some future date. But whereas in many societies such speculations have had an abstract, philosophical quality, for us they concern matters real and immediate. Our way of life changes so rapidly and continually that we think of change itself as a part of the routine of existence. We carry conceptions of the future as part of our stock of images and stereotypes, and we keep these conceptions up to date through reading, and reference to the mass media. We thereby sustain a stable market for the output of analysts, prophets and pundits, of all kinds and all qualities, who confidently inform us of where society is going, and why.

Some scepticism concerning this confidence is in order. The prediction of social change is a difficult business. History suggests that people are rarely able to discern the future development of their own society: their attempts to do so largely consist in a succession of disastrous misconceptions and profound failures of imagination. Even the immediate particular consequences of specific scientific and technological innovations have proved extremely hard to infer and describe: for every account offering successful predictions in this context there tends to be another, opposed account which represents a corresponding failure of prediction. Despite our continuing best efforts, and all the resources we are now able to bring to bear, the future remains elusive and difficult to grasp.

We have to speculate about future trends and tendencies of course, and in particular we need to think hard and long about the implications of continuing scientific and technological change. But speculation without sensitivity to its limitations can be worse than useless: inspired accounts of the coming age of the new technologies with their cornucopia of supposed benefits are engaging and seductive, but they can inspire nothing more than dangerous credulity unless we possess some sense of how to appraise them and assess their plausibility. For this reason I shall concentrate here on some of the general problems involved in understanding social change and the role of science and technology in engendering it. I shall discuss *how* we ought to think about this rather than *what* we ought to think, and avoid making too many empirical assumptions or specific predictions, which are all too liable to turn out to be wrong.

Inevitably, some minimal empirical assumptions must be made about likely future developments. It seems eminently reasonable, for example, to assume that scientific research and technological development will continue on an ever-increasing scale, at least in the industrialized countries. What we have seen occurring over the greater part of this century seems bound to continue throughout the next. Our stock of instrumentally applicable knowledge and technical skills and competences will continue to proliferate, and the stream of innovation will continue unabated with consequences both for the economy and system of production, and for society generally and our total way of life. Scientific, or scientistic, ways of thought, and forms of justification and legitimation, are likely to become even more entrenched and influential. The great institutions which both harbour and exploit scientific and technical expertise are likely to retain their dominance – at least, no discernible threat to it is anywhere apparent at the present time. The central question would appear to be: what will be the consequences of this continuing dominance of science and technology and their carrier institutions; and by what processes or mechanisms will they be brought into existence?

Note that the question does not seek to identify the impact of science itself, separately. It does not ask where science is going to take us, or what the effects of scientific research alone will be. Such questions are, I think, misconceived. When we consider the present significance of technical advance it is necessary to think of science and technology acting together, together adding to our stock of knowledge.

In the past, the task of advancing knowledge was often assumed to be the exclusive preserve of science and scientists. Science discerned new aspects of the world, and technology put these discoveries to good use. Science was a matter of creative exploration, technology of mundane application. Technology uncovered nothing new but merely made use of what was already known. Technology was merely applied science. But this conception is completely wrong. Technology is not applied science, nor anything remotely similar. Nor is technological development necessarily dependent upon advances made in science. Certainly current technological development is not so dependent. Modern technologists and engineers possess their own expert knowledge, established in their own work situations, with their own procedures for passing it on from generation to generation. A good proportion of this knowledge is non-verbal, embedded in skills and competences, in diagrams and visual representations, in the very artefacts of technology themselves. But it is no less knowledge for being non-verbal knowledge. And as knowledge it serves as the basis for creative technological activity, for the production of new knowledge, new technique, new artefacts, and thus new technology, entirely out of the existing resources of technology itself. New technology may develop out of old technology, just as new science develops out of old science: new engines are typically developments and refinements of old engines; new circuits elaborations of old circuits; new instruments adaptations of old instruments.

It is true of course that science is useful to technologists, that they do seek to exploit whatever recent scientific advances seem relevant to their purposes, and that some fields of modern technology are very heavily dependent indeed on the recent results of basic science. But it is equally true that technology is useful to scientists, that scientists seek to exploit whatever technological advances seem relevant to their purposes, and that some fields of modern science, fundamental particle physics for example, astronomy and astrophysics, much of X-ray crystallography, are very heavily dependent indeed upon recent advanced technology. This does not tempt us to regard basic science simply as applied technology; and no more should we be tempted to regard technology simply as applied science. We must recognize that both scientific and technological activities make genuine, authentic and significant contributions to the growth of knowledge.

As we have grown to realize that the growth of knowledge is more than merely the advance of basic science, so our conception of knowledge itself has become more sophisticated. In the past, the advance of knowledge was thought of predominantly as an extension of perception to encompass new objects and entities lying beyond the range of immediate apprehension. Pure science, of which physics has always been the paradigm example, was thought to perform the task of discovering these entities and subordinating them to our understanding. Thereby our general understanding of reality was advanced, our awareness expanded. The way the world always and everywhere operated was made intelligible, the laws it always followed were brought to light. The stock image of basic science, and thus of knowledge generally, presented it primarily as a cosmology, a representation of the basic nature of the world. But today we no longer think exclusively of knowledge of new objects or new kinds of entity. True, we still have a healthy respect for fundamental particle physics and accord it a specially honoured position in the hierarchy of the sciences. But we are now much more aware that knowledge is a matter of knowing how, as well as of knowing what, and that the accumulation of know-how represents the advance of knowledge just as much as the accumulation of observations and items of information.

When we think of the frontiers of knowledge today and the most challenging and significant fields of research, we no longer think only of the most basic scientific fields, and of disinterested investigations. We think also of fields like genetic engineering and biotechnology, information theory and information technology, artificial intelligence research, micro-electronics, the science and technology of materials. These and many other important recent fields of work, concern themselves not with general aspects of the world but with specific ordered structures within it. They seek to understand the special features of these structures with a mind to predicting and controlling their operation in specific conditions and circumstances. They are concerned to produce knowledge which will guide and inform specific human actions, rather than knowledge generally valid as a representation of the basic nature of things. A stock image of knowledge as a cosmology or world view, which is of questionable adequacy of course, even when applied to traditional basic sciences, is simply out of the question as a representation of knowledge in these more recent fields: an image of research producing universally valid truths, and technological development finding uses for these truths, falls apart completely when confronted with the way in which these fields operate. In these fields research produces knowledge in a way directly conditioned by possible applications (just as indeed it has long been produced in fields such as metallurgy or medicine). Clearly, therefore, there is a need for a stereotype of knowledge which stresses its instrumental character; its manifestation as technique, competence, procedure; its intimate connections not so much with perception as with action. And indeed such a stereotype does seem slowly to be emerging.

It is difficult to say precisely how the present relationship between scientific and technological activities might best be described. Certainly any simple model, or brief account, is likely to prove unsatisfactory in some respects. Perhaps the best thing is to think of science and technology as still partially separate activities, albeit increasingly intertwined together and difficult to unravel, both involved in the creation of new knowledge and new competence, and both making use of resources provided by the other as they do so. Looked at from the outside science and technology then appear as interlocked, interacting parts of a larger system from which new techniques and competences emerge, new possibilities for innovation flow.

Not just science then, but science and technology, and indeed other innovative activities within the economy generally, are likely to continue to provide us with a copious stream of technical advances and developments. Our ability to act upon and control our physical environment will be continually enhanced, supported by more and more instrumental capacities. And the capacities will be adaptable to many purposes. They will serve as versatile, flexible resources for action, rather as a range of general purpose tools and instruments will serve as versatile, flexible resources for action. The question is, What will we do with all this instrumental power and capacity? What difference will our possession of it make to us?

I have no specific answers to give to this question. I do have my own hunches about future trends and tendencies, and my own beliefs about which technologies will be most profoundly involved in our future living and how. But I doubt that these hunches and beliefs are any better informed than the large number of well-known scenarios of the future already offered by other writers and commentators, or indeed than the guesswork the individual reader might undertake himself. I want to avoid discussion of particular visions of technological change and its social impact, and consider instead the general framework of thought from which such visions have to emerge; for it seems to me that this framework is in very many cases, indeed in most cases, seriously inadequate.[1]

The major basic flaw is much of our thinking about the impact of technology is that we think in detail only about the technology itself. We analyse the technology in detail. We project as best we can its likely future development and elaboration. We strain our imagination to the utmost, thinking of possible uses and applications for the technology, possible artefacts and processes in which it might be involved. And then careful detailed thinking ceases. We assume that what is technologically feasible will be put into effect and that society will be landed with the consequences, like it or lump it. No explicit thought is given to the nature of society itself.

We tend to assume that society is inherently incapable of resisting the onrush of technological change and is impotent before it, but the standard grounds and reasons routinely put forward to justify this view are feeble in the extreme. Often the so-called reason is nothing more than a blind faith in the inevitability of 'progress' – with 'progress' being thought of as an external physical force, like gravity or magnetism perhaps, acting upon us and making us do this or that. Sometimes reference is made to the exigencies of international competition: 'if we don't do it then the Japanese will, and wipe us out.' This way of insisting that if it is possible it must be done is very popular with our industrialists and manufacturers, who often present keeping up with the competition not just as expedient but as absolutely necessary. Thus the industrialists of Europe, Japan and the United States offer protection to their populations, rather as rival gangs of thugs and hoodlums offer protection to restaurants, casinos and brothels; the same arguments are paramount in each case. But just as it is possible for single societies to carry on without the operation of protection rackets so also can the relations between societies be carried on without them. Unfettered international competition is surely not something given in the nature of things to ensure the triumph of the latest technology.

It is all too easy to find examples of the approach to social change which is obsessed by technology and indifferent to society. We know that immense possibilities have opened up in the realm of communications technology with the development of satellites, computerized information storage and transfer, and the micro-circuitry allowing sophisticated yet inexpensive personalized communication facilities to be produced on a vast scale. Accordingly, some commentators speak of our future as members of a 'global village', wherein all of us throughout the world will be involved in intense, immediate, personal communication with each other, and cultural barriers will weaken and be set aside. We know that automation is proceeding apace, and that with the development of sophisticated control systems it will become applicable to more and more complex processes; it will thereby become capable of replacing many more forms, and more skilled and sophisticated forms, of human labour, at ever higher and higher levels in the occupational hierarchy. Accordingly, it has been said that we are moving inexorably towards a world of leisure where only a small proportion of the population will be employed, and then only for a small proportion of their time, and where release from the demands of labour will create new opportunities and new ways of life. In the early 1960s it was realized that the use of new genetically selected strains of cereals could radically increase crop yields in the underdeveloped countries, and these strains were rapidly distributed and made available on a large scale. Accordingly, it was said that the consequent 'green revolution' would effectively put an end to the problem of hunger in much of the third world, and do away with the perennial outbreaks of famine and mass starvation hitherto so familiar.

It is not just optimistic accounts of the benefits of technological advance which manifest this tyle of argument: deeply pessimistic social commentators and critics of modern technology often set their case within much the same framework. Automation and robotics will, it is said, cause widespread unemployment and create a demoralized, unwanted underclass of idle citizens. Information storage systems and a developed technology of surveillance will, it is said, destroy privacy and individuality, and lead to a centralized, authoritarian, ultra-conformist, bureaucratic state. Advances in medicine and related sciences will, it is said, produce a population with an ever-increasing proportion of very old people, throwing a greater and greater burden upon the active and vigorous minority.

The simple accounts and arguments to which I have just very briefly alluded illustrate a common form of thinking sometimes known as 'technological determinism'. To be precise they illustrate a particularly strong and unsophisticated form of technological determinism which sees technological change as the cause and inspiration of social change generally, and which proceeds as if the latter can in some way be read off from the former. In some circumstances, with some kinds of social change and development, this may seem an extremely plausible and seductive approach. But it is quite fundamentally misconceived. It implies, without any justification, that people and their institutions are in some sense controlled by their own technology and somehow compelled to realize its inherent possibilities, so that one may speak of 'the implications' and 'the effects' of a technological advance without any regard for the social context in which it occurs.

This is simply false. One cannot read off social consequences directly from technological change. Technological change does not exert such a degree of coercion upon us. We are not, for example, obliged and compelled by our present military technology to proceed to the fighting of a nuclear war. It is up to us. (This particular example may serve to indicate why some people regard technological determinism as an altogether dangerous and pernicious way of thinking. On the other hand, given the present state of the world, some readers may find the example itself unconvincing. If this is so, think instead of how large stocks of chemical weapons were held by both sides at the beginning of the Second World War, yet were used by neither.)

Consider once more communications technology and the 'global village'. No doubt communications technology makes the global village a real possibility. But it is surely in no sense an 'implication' or necessary consequence of that technology. Is it not likely that the technology will be taken up and used differently in different kinds of society? Is it not possible that it will have different 'implications' in open societies like the United States, and more closed and secretive ones like Great Britain and the Soviet Union. Anyone familiar with the problems of direct dialling to Moscow at the present time may more readily be led to scepticism on the future realization of the 'global village'. The 'implications' even of current communications technology do not all seem readily to unfold themselves in the context of Soviet society.

Whatever approach we take to the study of social change and its relationship to technological advance, it is essential that insights into technical matters be matched and complemented by a properly detailed understanding of the social context. Technology is always a resource in the hands of people: it has no life of its own, but is developed and used under their control.[2] People acting collectively within the framework of the existing social institutions determine the impact of technological innovations. Indeed, they effectively determine what counts as a technological innovation rather than as a mere piece of idle ingenuity. Admittedly, there are cases where the use of a technical advance is so obvious that it hardly needs stating, where everyone agrees on the appropriate use, where there is no vested interest to make a case against that use. In such cases it is easy and straightforward to read off the use directly from the technique, or rather to imagine that this is what we are doing. In such cases we do not think explicitly about the social context, but simply take it for granted: if a drug selectively kills cancer cells its use and application may seem obvious and uncontroversial; likewise if a technique restores eyesight, or cures paralysis, or alleviates the effects of ageing. But cases such as these should not tempt anyone into the general habit of thinking that only technology matters, and that its 'implications' are bound to be realized.

Even in the most apparently straightforward and unproblematic circumstances the habit of thinking one-sidedly about technical advance can be disastrously misleading. Think again of the case of the 'green revolution' based upon new high-yielding cereal varieties. These varieties may seem quite unproblematically to be 'technological benefits', and their proper use may seem obvious and scarcely in need of detailed discussion. But there is nothing inherently beneficial about high-yielding cereals, or indeed about any form of high-yielding agriculture. The 'obvious' connection between 'high yield' and 'benefit' is simply a product of lazy thinking. On reflection most people would agree that what is desirable and 'beneficial' in this context is a satisfactory and stable relationship between people and their food supply, so that dire want does not exist, and famine and hunger become things of the past. No technical resource in itself can provide this. What is needed is a suitable relationship between people, their institutions and their techniques. But even though this is obvious enough to most people on reflection, the temptation to think purely of techniques and their intrinsic 'benefits' remains immensely strong, as does the temptation to respond to social problems purely and simply by looking about for some alleviating technique or technology. We are beginning to get used to the fact that social problems are rarely if ever solved in the long term simply by throwing technology at them, but it is a slow business.

It is worth asking why our thought should have this unfortunate tendency to concentrate upon the technical and to forget the social dimensions of change. Why are untenable forms of technological determinism, which treat technology as an independent power acting upon us and compelling our actions, so widespread? Perhaps it is because so many people in society experience the effects of technological change immediately and directly, but are not so aware of the distant decisions and strategies which produced these changes. New equipment may make work more flexible or interesting, or alternatively more fragmented, boring and repetitive. New techniques may reduce noise in the environment, or atmospheric pollution, or alternatively they may increase levels of radioactivity, or kill off or drive away many plant and animal species. Technology is clearly visible as the *immediate* cause of changes of this kind, and it is tempting to treat it as the *basic* or *underlying* cause of them as well, praising or blaming the technology itself for the effects of its use.

On the other side, of course, there will be the people who actually decided to introduce and use the technology to bring about the changes in question, perhaps because the changes had favourable effects on a profit and loss account, or perhaps because they furthered some political ends or objectives. People in this kind of position, members of elites who actually decide how to use and deploy technology, will presumably be well aware of how technology is subordinate to human ends and interests. Unlike those wholly on the receiving end of change they will not themselves be tempted to see technology as an impersonal dominating force. But it may nonetheless be in their interest to encourage this way of thinking. They may prefer people to think that they are dominated *by* technology, rather than that they are dominated *through* technology. If the contingent decisions of elites and leaders are generally seen as inevitable, as necessitated by technological change and 'progress', it may be so much the easier to implement and legitimate those decisions. If people blame disasters caused by technological change on technology itself, that may at least be preferable to the blame falling upon those who control it: people are always in search of scapegoats, and if technology can be made to serve as such then it surely will be. Thus, interests at higher levels of society may well conspire with limited perceptions at lower levels to encourage us to think of technology as a determinant of human actions, rather than a resource for human exploitation.

It is intriguing to notice that our tendency to think in the framework of a strong technological determinism is very far from being a uniform one. Consider the recent advances that have occurred in techniques for manipulating human eggs and embryos *in vitro,* with the possibilities for scientific research and technological exploitation they open up. In Britain these advances have been the subject of a prolonged enquiry by the Warnock Committee, which has considered them in a very broad framework and suggested possible areas for legislation relating to their use. Needless to say, recommendations put to this Committee have varied widely, and the recommendations it itself eventually put forward have been variously received. But the interesting thing is that there has been more or less complete agreement on the need to evaluate these advances openly and in the widest possible framework, taking account of technical considerations, public attitudes and opinions, political and economic interest, legal practice, moral and ethical arguments, religious sensibilities and religious doctrines. And there has been complete agreement too, that the use of these advances should be channelled and restricted to whatever extent and in whatever way is necessary to keep their effects within acceptable limits. There has been no question here of knuckling under to the 'implications' of technology; few doubts that the use of technology would be actively controlled and tailored to its social context.

I have not made mention of the Warnock Committee in order to recommend its proposals or even to advocate it as a desirable kind of response to the problems posed by technological development. We could, I am sure, do very much better. But the work of the Committee and the activities surrounding it show that, in some contexts, we routinely think of technology as a resource over which we have full control, and concerning which intensive and wide-ranging debate and discussion is important. What is puzzling is our failure to address all issues concerning the use of technology in this way. We seem to accept bizarre distinctions between different kinds of techniques and procedure. Apparently, the use and manipulation of the ovum and the blastocyst raises deep ethical and religious questions, whereas the use and manipulation of plutonium does not. Priests may pronounce upon the one, along with matters appertaining to abortion, contraception and the like, but not upon the other, or allied matters of military policy and industrial development. The immemorial doctrines of ethics and religion have authority in the one sphere, the imperatives of 'progress' and technical advance in another. The ovum and the blastocyst have a moral and ethical significance, whereas the complete human being would appear to have a merely political and technical one.

Perhaps the ovum and the blastocyst have been thrown to bishops and theologians as a kind of consolation prize. Having become largely redundant in modern societies as commentators on the central business of life, they are offered some authority over what comes immediately before and what comes immediately after, as a kind of sop. For myself, I confess to no disquiet at seeing the representatives of religion effectively deprived of influence on all matters of real substance in society in this way. But the notion that moral issues begin where living ends remains a distinctly peculiar one. And any notion that we should only seek to command and control a technology when it scarcely affects us is surely quite untenable.

IMPOSSIBILITIES

A continuing stream of new knowledge, of inventions and innovations, may increase the possibility of solving our perennial technical, manipulative problems, but it does not provide answers to moral and evaluative problems. It is generally recognized that the results of scientific research may have the one kind of utility but not the other. It is generally recognized, too, that scientific modes of reasoning and general styles of thought are not a sufficient basis for the solution of moral and evaluative problems. Science is concerned with empirical matters. It claims to tell us what is, but does not pretend to tell us what ought to be. And there is no logical procedure to carry us securely from a statement about what is to a statement about what ought to be. Whether it is thought of as a body of knowledge, therefore, or as a method of enquiry and inference, science cannot resolve moral and evaluative issues. This, at least, is the generally accepted view.

But for all that it is widely believed that science is morally and evaluatively neutral, it is also widely believed that science plays the crucial part in the actual solution of moral and evaluative problems. The claim is made again and again, both in serious appreciations of science and its social significance and in the polemical tracts of scientistic philosophers. And informal convictions to much the same effect are widely held by ordinary people and by natural scientists themselves. Quite rightly, it is held that science can *inform* evaluative decisions, and ensure that they are based upon sound empirical inferences. Science may provide the best possible context for the taking of an evaluative decision. But, more questionably, it is often also assumed that once such a decision is properly informed and properly considered, everyone involved will agree upon how it should be made. The assumption is not that the 'is' provided by science will logically compel a particular 'ought', but that the 'is' will naturally lead society, as a matter of fact, to one particular 'ought'.

This dangerously optimistic assumption tends to be justified in two ways. It is sometimes thought that we all of us share some common basic moral sensibility, some inherent conception of justice and fair dealing, so that once we understand a situation empirically we will naturally tend to agree upon what ought to be done in it. More commonly, and more plausibly, it is assumed that the members of any given society share some common values or principles, on the basis of which they take moral and evaluative decisions. Because these principles are agreed upon, concrete decisions will also readily be agreed upon once the correct empirical information relevant to the decisions is provided. Thus, in modern democracies the assumption is that the needs and preferences of all citizens are equally important and should carry equal weight when decisions and judgements are made. Legitimate collective decisions will, it is assumed, reflect social wants and preferences which are themselves true reflections of the wants and preferences of the individual members of society.

This picture of how 'society' operates is simple in the extreme. But we do often routinely think of society in such very simple-minded terms. When we think of a society accepting a policy or taking a decision we recognize the difficulty and complexity of the business of gathering the relevant information, but we tend to regard the matter of acting on the information as simple and straightforward. It is merely a matter of what society wants, we tend to think, or of which of the feasible alternatives it prefers. Once empirical, scientific procedures have sorted out the facts, the evaluation of the situation will follow automatically.

Here is an attractive account of the relationship between scientific knowledge and evaluative judgements, and hence, more generally, of the relationship between science and its social and political context. Science provides the facts and empirical inferences relevant to a decision or judgement; society has its wants and preferences; the facts show which decisions will satisfy which wants and preferences; thus science, in providing the facts, effectively determines which decision society will take. This account, however, attractive as it is, is quite misconceived. It glosses over the social processes which are always involved in the collective taking of decisions. It focuses attention on the role of the facts, the scientific evidence, at the expense of the people who take account of the facts, and the social relationships in which these people are involved. If we wish to have a properly adequate understanding of the relationship between (scientific) fact and (social) evaluations, and it is most important that we develop such an understanding, then we must broaden the perspective and become curious about what collective evaluative decisions actually involve. Unfortunately, as we do this we encounter limits on the relevance of empirical knowledge and scientific argument. We find social problems which must be solved, but which are impossible of solution within a purely scientific framework. We have to accept some tempering of any optimism about the beneficial effects of science within the social and political sphere.

Consider again the view that society has its wants and preferences, and that science indicates how they might best be satisfied. What is meant by the claim that society has a want or a preference? We are reasonably clear how an individual might prefer one thing to another. But how is it possible for a society to do so? A society, after all, is made up of many individuals, typically with different characteristics, in different situations, and therefore typically having different preferences. How can a society of this kind have 'a' preference, or indeed have any general characteristic at all? Is it not a mistake to speak of 'society' in this way, as though it is an individual writ large. Should not this informal habit of speaking be recognized as incorrect, and set aside?

Most of us would probably agree that it is wrong to think of society as an individual writ large, with wants, preferences, needs and other typically individual characteristics. But most of us would probably wish to continue to speak of 'society', nonetheless. (There are many references to 'society' in this book.) What some of us would perhaps say is that society is indeed really a collection of a large number of separate individuals, not a single individual entity, but that we talk of the preferences etc. of society as a kind of shorthand for individual preferences. The individuals preferences of members of society 'add up' as it were, and allow us to speak of overall social preferences; and so similarly with wants, needs and the rest. Unfortunately, all this would be wrong. A society cannot be satisfactorily described as so many separate individuals with independent preferences. And even where separate individual preferences exist in a society, they cannot be satisfactorily 'added up' to yield an indication of 'what society prefers'. This whole imagery for thinking about society, a commonly used, taken for granted imagery, is inadequate and misleading.

To see how this is so let us think a little more concretely about the wants and preferences of people, and how they can be satisfied. Many of the preferences of individuals in society can of course be treated at the individual level, and not combined or added together at all. If some people prefer red wallpaper and others blue wallpaper, then red can be provided for the one group of people and blue for the other. Preferences of this kind are easy to deal with, and easy to talk about. But not all preferences are of this kind. If some individuals would like to have a nuclear reactor round the corner and others would not, it can be difficult to satisfy both groups at once. If some people prefer a Conservative government, others Labour, others again Alliance, it is not possible to satisfy them all. It is when dealing with problems of this kind that we are tempted to speak of preference in general terms, indeed, what else is it feasible to do? We refer to a social preference for something, or a collective preference, or an overall preference. At least, we are apt to think, if we cannot satisfy every individual preference we can satisfy a general social preference, we can seek to provide what is most preferred by society as a whole.

Certainly, this way of thinking is important in democratic societies such as our own. Governments like to present their policies as the socially preferred policies whenever they are able, and to present themselves as holding a 'mandate' from society. And they like to present their opponents as ignorant of 'popular feeling' and 'public opinion'. Needless to say, this sort of political rhetoric is not to be taken too seriously. Most of the present (1984) British House of Commons seem well content with an electoral system which, in theory, could produce a parliamentary majority from 5 to 10 per cent of the recorded vote or even less. And most of them are certainly content to enjoy a parliamentary majority based on what was very much a minority of votes cast at the 1983 election. For practising politicians a good electoral system tends to be one which allows them to win and to exercise power. But it is nonetheless important, in our society, to be able to justify decisions as in accordance with social preference, with 'what people want'. Our ideals encourage us to think in this way and to some extent to judge our politicians in this way. And politicians know it and take account of it. Where it is expedient they use surveys and public opinion polls to show how their policies express social preferences. They confidently point to the aggregate of individual responses recorded by these polls, as though they transparently indicate social preferences.

Such confidence is, of course, a part of the good politician's professional image: he is generally equally adept at affecting a profound scepticism about those opinion polls which appear to be against him. The aggregation of the individual responses produced by polls of this kind is known to be full of difficulty, and the misinterpretation or skilful manipulation of such materials can be refined to a fine art. An artful defender of coeducation will be careful to point out that most parents would like to send their sons to coeducational schools, and not to single-sex institutions. This apparently is true enough. But it seems to be equally true that most parents would like to send their daughters to girls-only schools. There seems to be a social preference for male-only coeducation, a preference which is difficult to satisfy. Similarly, an artful defender of the United Kingdom and its continuation will be careful to emphasize that most of the citizens of Northern Ireland wish the UK to persist, and are in favour of the Union. No doubt this is true enough. But the case of his critic will be that most of the people of Ireland are against the Union, and would wish to see the UK dismantled, given that it could be said to exist in the first place. Should we speak here of a social preference among the Northern Irish for the Union, or of a social preference among the Irish for national independence? What is the 'appropriate' way of aggregating individual opinions?

These of course are mere technical kinks. They illustrate only that to talk about a social preference one must be clear about precisely what the relevant society is, and precisely what is preferred by the individuals in it. The real problem, the deeply interesting problem, is this: Can we, in ideal circumstances, reliably convert *individual* preferences into a consistently well behaved and sensible measure of *social* preference? Can what the politician artfully does be done properly? Can our knowledge of 'what people want' individually be used to give a reliable indication of 'what people want' generally. The answer to all these questions is negative. There is no specific method of calculating social preference which we would find consistently satisfactory and well behaved. There is no algorithm for converting individual preferences into a social preference which we could feed into a computer and then routinely trust and forget about.[3]

A concrete example is the best route to an understanding of this point. Consider the problem of producing a voting system which makes it possible for members of a society reliably to elect the socially preferred political party into government. Imagine, for simplicity, that there are but three parties, and that the system must use the individual preferences of voters to produce a governing party (the socially preferred party) from these three. This is a handy example, since we can use the British General Election of 1983 as an embodiment of it. In that election there were effectively three choices for voters: let us call them Thatcher (T), Jenkins (J), and Foot (F). But the electoral system was not one which would reliably yield a sensible social preference from the individual preferences fed into it. Our electoral system might, for example, allow the least individually preferred option to become the government. It is perfectly capable, in some circumstances, of converting T 30 per cent, J 35 per cent, F 35 per cent into T wins. Indeed, far worse excesses are possible. We can scarcely call a system capable of working like this a reliable indicator of social preference. But could we improve the system, or substitute another, so that social preference was reliably indicated?

Imagine that a system of simple majority voting is introduced. Voters are offered a choice of two options – say, T versus J – and the winner on that vote is pitted against the third candidate, F. This at least stops any party which a majority of voters consider the least attractive. And it does seem to work reasonably well, intuitively speaking, most of the time. But consider the situation outlined in table 5.1.

Table 5.1

Groups of voters (all less than 50% of the voters)	*Order of preference*		
	1st	*2nd*	*3rd*
A	T	J	F
B	J	F	T
C	F	T	J

Here there are just three kinds of order of preference, but no order is favoured by more than half the voters. Group A voters all individually prefer T to J, and J to F; in group B the order is J → F → T; in group C it is F → T → J (I shall use → to mean 'is preferred to'). Note how realistic this case is. It is not all that far from what appertained at the election. Imagine that group A are Conservatives to whom the Labour Party are anathema, and the Alliance at least tolerable. Group B are Social Democrats with a lingering affection for their old allegiance, the Labour Party. The third group are socialists, for whom anything at all is preferable to a vote for the traitor Jenkins.

Suppose now that we use our system of simple majority voting to see whether Thatcher, Jenkins or Foot is the socially preferred option. (It will help at this point, and for the rest of the section, to have a pencil and paper handy.) We take two options, say Thatcher and Jenkins, and find that Thatcher is preferred. (Groups A and C combine to outvote B.) Then we set Thatcher against Foot, and find that Foot is the overall winner. (Groups B and C combine to outvote A.) So Foot emerges as the social preference. But in fact a clear majority of the voters actually prefers Jenkins to Foot! (Groups A and B will outvote C.)

Looking at the situation in more detail, what we find is that if simple majority preference is taken as the 'social preference' of the electorate, then Thatcher is socially preferred to Jenkins, who is preferred to Foot, who is preferred to Thatcher:

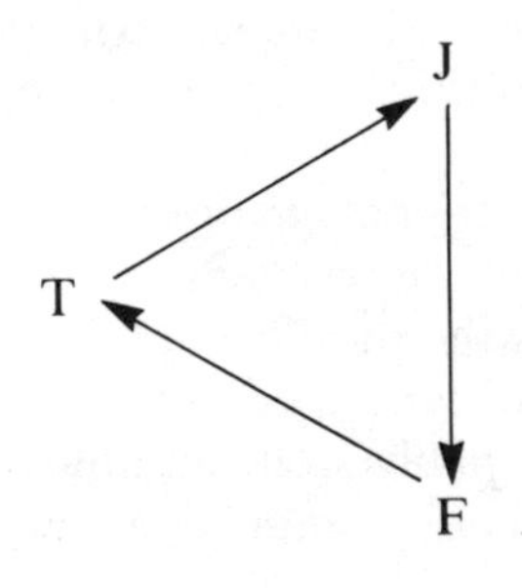

The 'social preference' is cyclic. And in these circumstances the system of double pairwise voting always gives the verdict to the last option involved. In the case cited here, Foot won because he entered the voting last. If Thatcher had entered last, she would have won; if Jenkins had been left until last, he would have won. This is easily confirmed with pencil and paper. Thus the winner in these conditions the 'socially preferred' choice for government, is entirely an accident of how the voting is carried out: this is all it is to do with. Clearly we cannot rely on a voting procedure which is liable to produce purely accidental results of this kind, and presume to call such accidents 'social preferences'. Even professional politicians might evince a hint of a blush were they to refer to such an accident as evidence of 'a mandate from the electorate'. We must seek yet another system – one which is immune from such accidents.

Unfortunately, there can be no such system; even in this very simple case there is no way of ordering a voting system so that it does not snarl up, in some circumstances, as the system above snarled up. *More generally, there is no sensible way of ordering any set of individual rank orderings, where three or more choices are involved, which does not similarly snarl up in some circumstances.* This claim can in fact be made very precisely and justified very convincingly: it is a part of what is implied by 'Arrow's Impossibility Theorem', a widely known and accepted result with considerable significance in economics and political science.[4] But there is no need for undue precision here, or for any examination of what is a difficult proof of the general result. It is sufficient to be aware that any intuitively sensible method of combining individual preferences to produce a 'social preference' will fail to satisfy in some conditions of use.

Why do snarl ups of this kind occur? Why cannot we aggregate the preferences of individuals without being liable to run into such snags? Basically, it is because of the different characteristics of individual preferences and aggregations of the preferences. Individual orders of preference are what is called *transitive.* They conform to a general rule: if A is preferred to B, and B to C, then A is preferred to C. Psychologically this is overwhelmingly plausible: if Thatcher is preferred to Jenkins and Jenkins to Foot, what individual will nonetheless prefer Foot to Thatcher? If wine is preferred to beer and beer to water, what individual would at the same time prefer water to wine? It would have to be an extremely strange individual. With such an individual one could perhaps exchange a bottle of wine for two bottles of beer, then the beer for four bottles of water, and then the water for eight bottles of wine, making a tidy gain on the cycle, and looking forward to bigger gains on the next cycle. Intransitive individual orders of preference would quickly lead to bankruptcy; in practice individual activity cannot be based on them; it is right to assume their non-existence. But 'orders of preference' produced by aggregating together individual preferences may very easily be intransitive, and go round in circles, without any implausible implications. And indeed all plausible ways of aggregating individual preferences together do produce intransitive, cyclic, orderings under some conditions: orderings of this kind are useless to us as 'social preferences'. After all, our underlying aim in seeking to talk of 'social preference' is to be able to provide 'what society prefers', to be able legitimately to treat society like an individual writ large. And an order of 'social preference' which is liable to go cyclic and produce purely accidental results, is simply too unlike an individual preference to count as a legitimate substitution for it.

Consider now the practical significance of all this. In making evaluative judgements and decisions we would wish to take proper account of 'what society prefers'. But it is impossible to construct a satisfactory measuring procedure to ascertain this. All actual measures of 'social' or 'overall' preference (or indeed of 'social' or 'overall' anything initially located in individuals), are unsatisfactory in some respects: all such measures are liable to favour some individuals over others in some circumstances. Therefore, when we look at the procedures actually used in a society for aggregating individual preferences, we must take great care to see them for what they are. They must be understood as the customs or conventions of the society, as 'institutionalized practices', the use of which has specific consequences. And the particular institutionalized practices of this kind which are routinely used in any society must be recognized as having important implications for the distribution of power in the society: they will be practices which, in imperfectly representing 'what people want', will preferentially help and assist particular individuals to get what they individually want.

It is most important that this point be recognized, since our society is replete with practices of this kind. The whole of our ordered, organized collective activity is organized around them. It is not just a matter of local and general elections. Ways of aggregating individual opinions are routine and taken for granted in company board meetings, in trade union meetings, in the informal operation of political parties, in legal contexts, in all the great bureaucracies: committees of all kinds in all contexts employ them; societies and associations of every variety find a use for them. And even as they are the necessary means through which individuals may express their views and urge their preferences, so are they also, invariably, the means through which some individuals gain priority over others.

Think again of the three by three voting situation discussed above, and imagine that the pairwise majority voting system is in use. Let us continue to think of the three choices as T, J and F, as before, even though we are no longer concerned with the specific case of the British election. Now much of the time the pairwise majority voting system will decide between three choices to the satisfaction of everyone. But we know that in some circumstances the awkward pattern

T J F
J F T
F T J

may arise. When it does, the formal inadequacy of the voting procedure has a practical consequence of great importance: those responsible for the voting procedure are able to influence which option wins the vote.

Suppose that those responsible for the vote wish T to win. They are obliged to apply the voting rules correctly, or the result may be challenged. But these rules allow them to set J against F, and then the winner of that pair against T, thus guaranteeing victory for T. Analogously, if they favoured J, they would leave J to the last vote, and similarly with F. More generally, in any size of system with majority voting those responsible for the voting procedure will confer advantage on their own favoured selection if they delay its involvement in the voting as long as possible: remember that the three by three case is being used to illustrate a general point.

In the last analysis, therefore, any voting procedure has to be thought of as a social institution, a routinized pattern of accepted activities. And this pattern, whatever it is, must be of a kind which creates an unevenness between people, which gives some people more rights and more powers than others.[5] Needless to say, some patterns of relationship between people are much more unequal than others, and many societies have established patterns which minimize the extent of the inequality. But even in such maximally democratic societies it remains impossible to understand decision and policy as some direct reflection of the aggregate of independent individual preferences; it remains essential to look to the routinized patterns, the institutions, in which preferences are ordered and expressed.

In general, it is a profound mistake to think of society in a simplistic way which takes no account of its internal structure, of the complex patterned interactions between its members and the institutionalized practices and procedures they carry out and sustain. It is tempting to ask what response 'society' will make to a given body of evidence or empirical information, or how 'society' will evaluate an innovation or a new technical procedure, but the temptation is best resisted. Society is not an individual writ large which can respond as one thing, a single undifferentiated whole. It is also tempting to move to the opposite extreme, to think of society as so many separate, autonomous individuals, each of whom can appraise scientific evidence, or evaluate scientific innovations, in isolation, and make his own independent response to them. But again the temptation is best set aside. The response of a society to scientific evidence or scientific innovation is embodied in a small number of decisions and judgements which have to serve for everyone. And how these decisions are made will always be a reflection of the institutional structure of society. This structure may allow a wide range of individual preferences to have a bearing on decisions, or it may not. But its routines will never function merely as the means whereby to 'add up' individual preferences: it will always be imprecise to speak of decisions made by the use of these routines simply as expressions of 'what society prefers'.

Necessarily, then, the resources of science are indirectly coupled to individual wants and preferences. It is not possible to design a direct coupling, a coupling which is, as it were, transparent to the preferences which must pass through it. Thus it is not true that in a given context of wants and preferences, judgements and decisions can be effectively determined by scientific evidence and scientific reasoning. The institutional structure, the established routines which connect individual wants and preferences into decision-making systems, will always have to be taken into account as well. Needless to say, this is one reason why the use and exploitation of science to satisfy individual wants and needs is often much less vigorously pressed forward than might seem appropriate.

DANGERS

In the preceding sections I have laid great emphasis upon the role of the social and institutional context within which science and technology develop. I have tried to show how attention to this context exposes the inadequacy of strong technological determinism, and demonstrates the need to think of technological change not as a compelling force but as an increasing supply of resources for our use. And I have tried to show too how attention to this context is essential to a proper understanding of the relationship between scientific evidence and collective evaluative judgements and decisions. I now want to continue with the same theme, by showing how close attention to context is essential in understanding the outcome of rational calculations made in the context. Most of us are familiar with the nature of rational calculation, but we tend not to think of it as an activity in a society, and we consequently overlook interesting features of it as a social phenomenon. Similarly, most of us are immensely optimistic about the power and reliability of rational calculation, but are unfamiliar with the dangers which may attach to it in a social context. We need to recognize that in a social context rational calculation can *generate* problems, and thus that even as we attack existing problems in society rationally and scientifically, we would be wise constantly to retrace our steps and consider whether any further problems might not be emerging in our wake. Rational calculation in society is best accompanied by a degree of reflectiveness: it needs to be seen as part of the problem as well as the means of solving it.

Our paradigm notion of rationality, just like our paradigm notion of preference, is linked to our understanding of independent individual behaviour. A person's rationality is his reasoning capabilities: it is manifested in calculation, in planning and prediction, in the working out of maximally effective courses of action, in inferring the best means to given ends. In all these tasks rational procedures, procedures of secure reasoning to the correct or most probable answer, are routinely used. There is actually considerable disagreement as to what precisely these rational procedures consist in, and how potent they are: some of the key problems in the philosophy of science concern these issues. But about one central aspect of individual rationality everyone seems to be completely agreed: it is never second-best.

Imagine someone at a casino, who describes the most rational method of placing bets at the roulette wheel, but then goes on to say that in fact he makes no use of this method, but actually employs another method which works better. What we would want to say here is that, if the latter method is the more effective, then it is also the rational choice. It cannot be that one method is the rational choice, but another the more effective. In our habitual frame of reference, rational calculation and rational inference cannot betray us. If an individual seeks to decide the best available means to a given end or goal, there can be no better way of deciding than the rational way: the best choice is the rational choice. If a better means subsequently emerges it immediately becomes the rational choice from the point of its emergence, replacing the old rational choice, simply because it is better. In our thought about individuals, and their choice of means to a given end, 'best' and 'rational' are near synonyms: they are not merely associated together empirically; they are locked together in the structure of our thinking.

If we are to understand the role of rational calculation in groups or societies, however, it is most important that these terms, 'best' and 'rational' are unlocked, that the connection between them is severed. What is perhaps the most basic intuition of our usual individualistic way of thinking about rationality has to be set aside. It is almost invariably correct to say of an isolated individual that, in as far as his own individual interest is concerned, he does best to act on the basis of a rational calculation of how that interest is best served. But it is not at all correct to say this of a group: it is not correct to say that in so far as their individual interests are concerned people do best to act on the basis of a rational calculation of those interests. This claim is for most of us very counter-intuitive. How can individual ends not be best furthered by the actions which are rationally and correctly calculated to be the best means of furthering them? If the ends are not best furthered, then surely the actions have not been rationally calculated. To suggest that it can be against people's best interests to reason soundly and calculate correctly sounds like a contradiction in terms.

Precisely because the point is so counter-intuitive, it is best to come to grips with it through a specific illustration. The one customarily employed is the story of the prisoners' dilemma. We must imagine two masked gunmen robbing a bank of a substantial sum, and making their getaway in a stolen car with the police in hot pursuit. As the police close in and arrest becomes inevitable, the gunmen manage to dump their loot over a bridge into a river. They surrender to the police with the reasonable hope of serving only a trivial sentence for the theft of a car, since evidence of the more serious crime is no longer available. Little do they realize that the police will shortly make an implacably logical appeal to their own rational self-interest, in order to effect their downfall.

Let us call the two prisoners A and B: both are in identical situations, so it suffices to look at only one of them, say, A. Naturally, the police are interested in securing a full confession from A, and seek to persuade him of the benefits of such an action. If his confession is used to convict his partner in crime, B, then, say the police, A himself will go scot-free (as a reward for turning 'Queen's evidence'). On the other hand, if B also confesses, and A's evidence is not actually needed to convict B, A's confession will still count in mitigation of his crime to some extent, and a 21 year sentence will be reduced to one of 20 years. Against this, A has to weigh the fact that if he does not confess he will not go scot-free, but will face at least one year in prison for car theft. He is also aware that his partner, B, is in the same position as himself, and that B's confession may perhaps be used to send him down for 21 years. All possible outcomes are summarized in table 5.2, and all the information in the table is known to A.

Imagine now that A is concerned solely with his own self-interest. He must seek to work out the best way of furthering that interest given his knowledge as set down in the table. His task is a very easy one: he can

Table 5.2

A's Sentence	*A's Policy*	*B's Policy*	*B's Sentence*
20	Confess	Confess	20
21	Not confess	Confess	0
0	Confess	Not confess	21
1	Not confess	Not confess	1

deduce his best policy with complete certainty. A glance at the top half of the table reveals that if B confesses, then A should confess also: a sentence of 20 years is preferable to one of 21 years. A glance at the bottom half of the table reveals that if B does not confess, then A should confess: going scot-free is much to be preferred to a year in prison. But B can only confess or not confess. Hence, *whatever B does, A does best to confess*. Needless to say, since B is in an identical situation to A, his rational calculations must take the same form. The same inexorable logic must lead B to confess whatever A does. So long as both prisoners are separately moved by the desire to remain in prison for as short a time as possible, so long as they are able to rely upon the knowledge in the table, and so long as they calculate rationally and logically, the outcome must be that both prisoners confess to the bank robbery.

Let us now consider how successful this reasoning on the part of A and B is in furthering their individual ends of keeping out of prison. By impeccable logical procedure and sound calculation they manage to limit their period in prison to 20 years each. What sort of an achievement is this? We can get some sense of it by making a comparison with alternative strategies. If A and B had had a dogged faith in irrationality so that they always did the unreasonable thing, then neither of them would have confessed and both would have facd one year in prison. Each would have made an individual gain of 19 years from irrationality. If A and B had both been lazy and had simply tossed coins to decide what to do, then they would have given themselves, on the average as it were, ten and a half years each – an individual gain of nine and a half years from laziness. Broadly speaking, as far as their individual best interests are concerned, rationality is much the worst approach for them both to adopt, indolence is considerably better, but irrationality is much the best. If only both had acted against their individual best interests they would have furthered those interests very effectively.

Table 5.3

→ *B* / ↓ *A*	*Confess*	*Not confess*
Confess	20/20	0/21
Not confess	21/0	1/1

For the record, rational calculation also produces the worst possible *overall* result, the outcome which is collectively the most adverse: a total of 40 years in jail is incurred by the two prisoners considered jointly, far and away the largest combined total, whereas irrational action by A and B gives the best obtainable joint total of two years.

The prisoners' dilemma graphically illustrates how the rational pursuit of individual objectives, in groups, may be self-defeating. But, of course, it is only an illustration. Let us try to get at the general form implicit in the illustration. First, consider the relevant data as re-arranged in table 5.3; the outcomes are written with that for A first and that for B second.

Clearly, the four possible sentences of 0, 1, 20, 21 merely represent any four outcomes for which a fixed order of preference exists: 1st (0), 2nd (1), 3rd (20), 4th (21). And the matter of confessing or not merely represents the making of any decision, any choice between X and not X, which affects the outcomes. So the general form of table 5.3 can be expressed more clearly in table 5.4, where the basics of what is involved are clearly perceptible.

A is in a position to move the overall outcome up and down in the table by choosing X or not X. But a move up is a move from A's 4th preference to 3rd, or from his 2nd preference to 1st. It is always rational. B can move the overall outcome from left to right. But a move to the left always obtains B a higher preference, and is always rational. So rational calculation by both A and B moves the final outcome upwards and to the left, giving each individual his 3rd preference. On the other hand, if both A and B had been irrational this would have moved the outcome down and to the right, and each would have obtained his 2nd preference. Because A and B were rationally seeking their goals in the same situation, their rational activity *combined* to produce a disappointing outcome for both of them. This is perhaps the crucial thing to remember, that perfectly rational calculations by two

Table 5.4

A ↓ \ B →	*X*	*not X*
X	3,3	1,4
not X	4,1	2,2

people combined to produce a worse result for both than if neither had been rational in the first place.

Notice also that if A and B are rational, are fully competent reasoners and calculators, they cannot at all easily escape from the fate to which their rationality dooms then. A 'no confession' agreement, for example, would not help. It would be irrational to keep it. To keep it would simply produce a lower preference than might otherwise be had.

For an isolated individual his reasoning capabilities are an asset: they are a marvellous resource in the solving of problems. For more than one individual, however, for a group or a society, these same reasoning capabilities may be a *source* of problems, and their possession something of a mixed blessing. Once cannot casually assume of a society that the more rational its members are, the better: one certainly cannot casually assume that where a society has problems, an appeal to reason is the way to an answer. *Many of our most important practical social problems may be thought of as unfortunate interactions of individual rational calculations.* Indeed our whole society is shot through with problems which can be interpreted in this way – as rational calculation snarling up because of the given, predetermined social context in which it occurs. (Note, however, and this is very important, that the calculators must believe that the social context is fixed and predetermined. What happens if they do not believe this? What happens if the prisoners try to bribe the police? What is rational calculation for the policeman?)

So long as people believe in their situations as given and fixed, there are many instances where their calculations tend to produce the prisoners' dilemma pattern. Calculations related to national security are a depressing example. Imagine that scientific and technological advance makes possible the construction of a new weapons system. For two potential adversaries this may present an awkward dilemma. Each opposed country may see the system as a valuable addition to its security, provided that the other country does not build it also. Each country may also perceive that it is better for neither to build it than for both: if both build the weapon, both incur large costs and both are left no better off than before in terms of relative military strength. Nonetheless rational calculations of individual self-interest would tend to press both countries into building the weapon, so that both waste their resources for no benefit. The form of the situation is precisely that of table 5.4 where A and B are the two countries, and X is the weapons system in question. If B does not build the weapon, then to build it is good for A; if B does build it, then A absolutely has to build it; either way it should be built. This is the argument.

Notice too that an argument of this kind may surround the development and realization of any scientific innovation at all. A new pesticide or a new fertilizer may increase a farmer's crop yield, but pollute rivers and waterways when widely used in large quantities. It may well be that the extremely unpleasant consequences of the wide use of the pesticide are such that even the farmer himself finds them an unacceptable price to pay for his gains. But the farmer would nonetheless be caught by the argument to self-interest. If nobody else uses the pesticide, he should use it since he obtains more crops and incurs no loss; if everybody else uses the pesticide he suffers the consequences anyway, and his using it gives him the benefit of more crops without making those consequences tangibly worse. Thus most farmers may well decide to follow self-interest and use the pesticide – to their mutual detriment.

The trouble here is that the benefits of the pesticide are *divisible,* but its drawbacks are not. The farmer can take all the benefits of the pesticide he buys to himself, as increased yields of *his* crops, and ensure that no benefit spills over to others. But the drawbacks of the pesticide are *indivisible.* They are shared around indiscriminately amongst everybody as environmental pollution. Only a small part of the adverse consequences of the farmer's own individual behaviour is experienced by the farmer. Thus the total adverse consequences of the farmer's own individual behaviour can far outweigh the benefits without the farmer having any incentive to desist. But as a member of a farming community all in the same position, he will nonetheless experience overall adverse consequences far outweighing overall benefits. He and his community are trapped in a prisoner's dilemma kind of situation. As individuals it is in their interest to use pesticide whatever anyone else does. As a community it is against their interest to use pesticide, and every individual in the community loses thereby.

Where a scientific innovation or a new product, when individually used, or a policy, when individually followed, confers *divisible* benefits which can be confined to individuals, but *indivisible* costs or disadvantages which cannot be so confined, it is liable to be taken up even if it

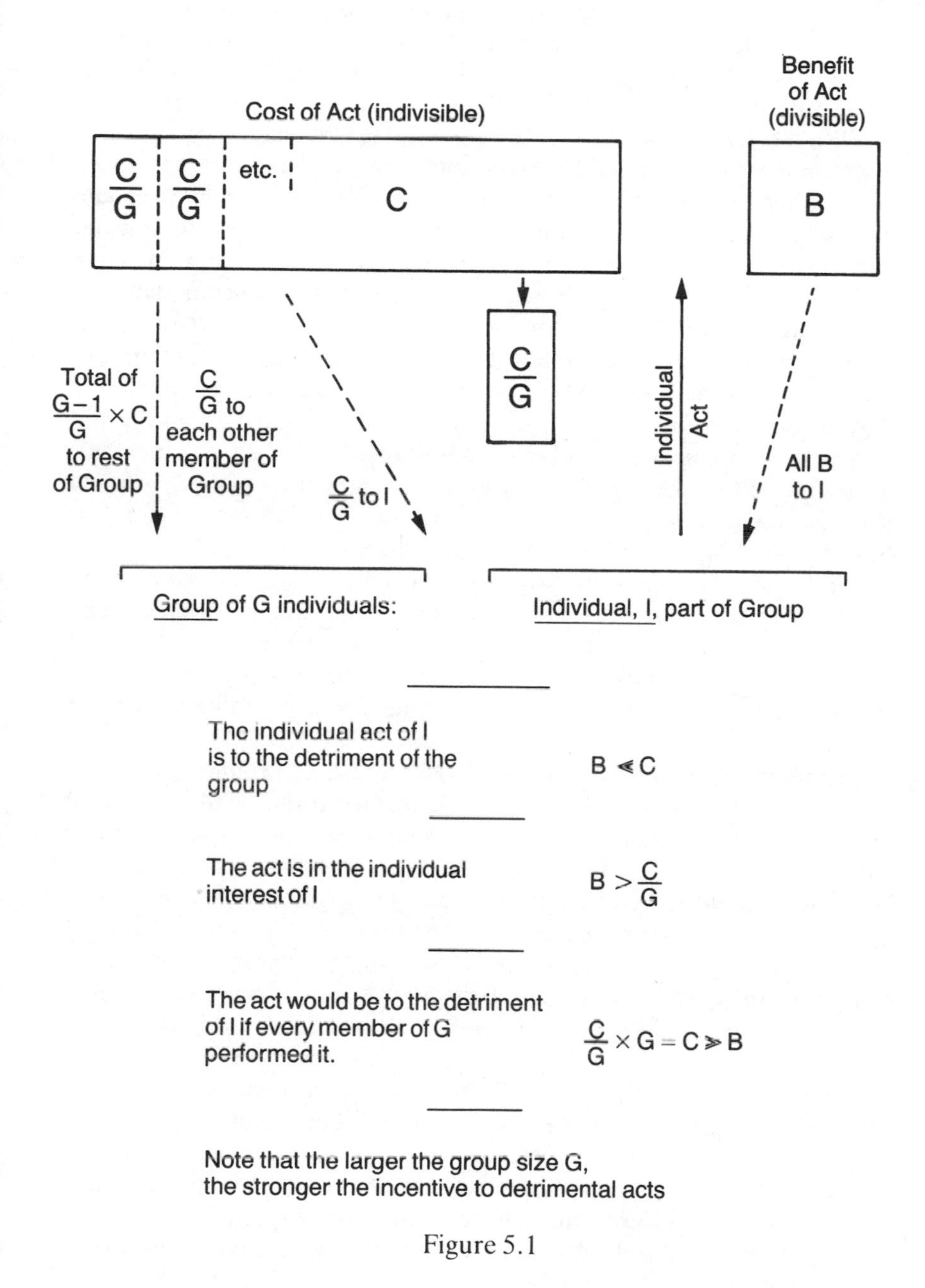
Cost of Act (indivisible)
C/G
C/G
etc.
C
Benefit of Act (divisible)
B
C/G
Total of (G−1)/G × C to rest of Group
C/G to each other member of Group
C/G to I
Individual Act
All B to I
Group of G individuals:
Individual, I, part of Group
The individual act of I is to the detriment of the group
B ⋘ C
The act is in the individual interest of I
B > C/G
The act would be to the detriment of I if every member of G performed it.
C/G × G = C ⋙ B
Note that the larger the group size G, the stronger the incentive to detrimental acts

Figure 5.1

actually harms those who feel impelled to adopt it (see Fig 5.1). Thus, merely to say that most people would wish to use an innovation if it became available is no argument at all in its favour. However one tries to work out the benefits of a scientific innovation, it should not be simply by adding up individual attitudes to it.

Prisoners' dilemma kinds of argument may not only encourage the performance of detrimental actions; they may also inhibit the performance of beneficial ones. As I write there is a water shortage in the south west of England, and its people are being exhorted to halve their water consumption. Such restraint now will be very much better than the enforced use of standpipes which will otherwise have to come later, as politicians are very rightly emphasizing over the media. Restraint is indeed in everyone's best interest. But unfortunately, not their own personal individual restraint. If everyone else exercises restraint, then any given individual can be unrestrained now, and yet also avoid rationing later; if nobody else exercises restraint, then one person's restraint will make no difference anyway. Hence, nobody should exercise restraint out of rational self-interest. To follow rational self-interest should inexorably lead to water rationing.

Where a policy when individually followed, or an innovation when individually used, involves *divisible* costs or disadvantages which can be confined to those individuals, but produces public benefits which cannot be so confined, its adoption is likely to be discouraged by considerations of individual self-interest, to the detriment of the interests of the individuals involved (Fig. 5.2).

Scientific research itself involves divisible costs to produce indivisible benefits. Since the results of basic science are traditionally published openly and made available to all, why should anyone in particular pay for it? It has often been argued in Britain that since basic research has so often turned out to be of enormous economic significance, it should be vigorously supported by government. But just as it stands this kind of argument is unsound. Why not let other countries do the pure science and incur the costs thereof? The British government could then simply support the exploitation of this science. Can Britain afford to be putting valuable resources and talented manpower into basic scientific research, when it could leave that to be done elsewhere and use its own scarce resources simply to exploit such research? Some small proportion of Japan's economic success is occasionally attributed to a parasitic policy of this kind, letting other countries do the fundamental work and putting its best effort into development and exploitation. So the argument is not a purely abstract one. But consider the consequences of every country's being moved by such an argument. Then there would be no basic research at all.

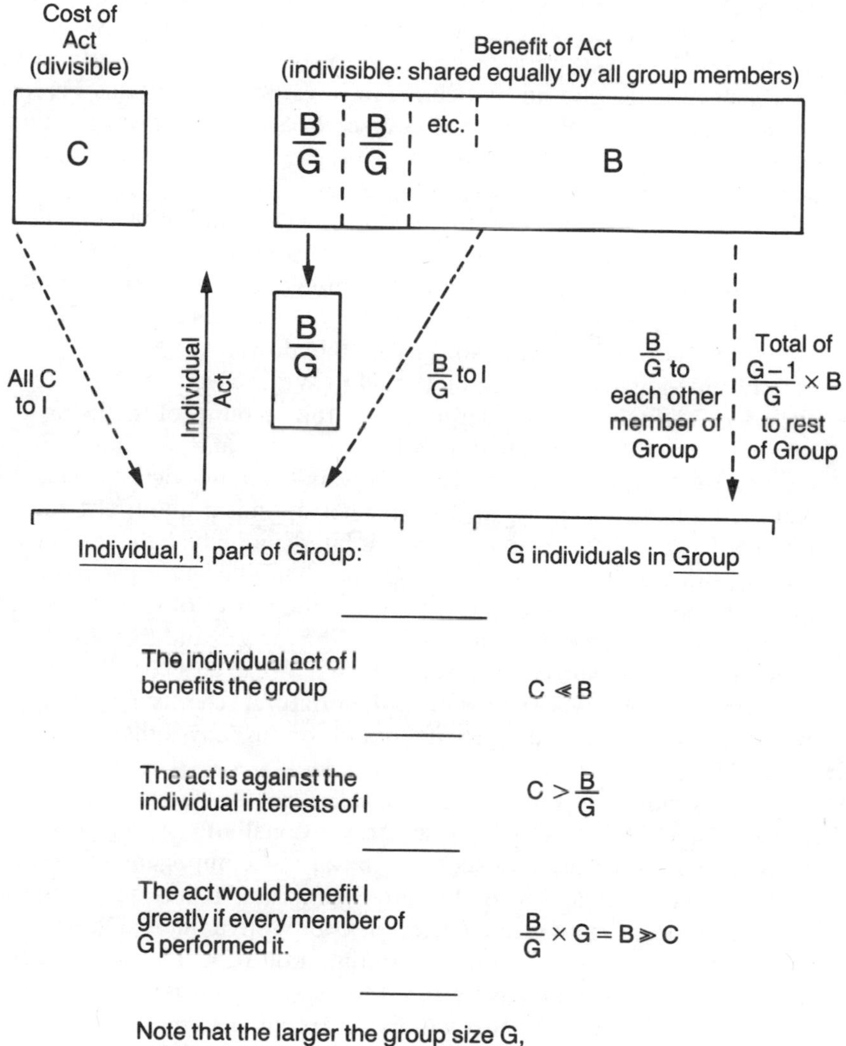
Cost of
Act
(divisible)
C
Benefit of Act
(indivisible: shared equally by all group members)
B/G
B/G
etc.
B
B/G
All C
to I
Individual
Act
B/G to I
B/G to
each other
member of
Group
Total of
(G−1)/G × B
to rest
of Group
Individual, I, part of Group:
G individuals in Group
The individual act of I
benefits the group
C ≪ B
The act is against the
individual interests of I
C > B/G
The act would benefit I
greatly if every member of
G performed it.
B/G × G = B ≫ C
Note that the larger the group size G,
the stronger the disincentive to beneficial acts

Figure 5.2

To understand the consequences of rational calculations made by members of groups or societies is a very difficult matter, involving a whole series of further ramifications which I shall not even touch upon here. It is enough if I have created an awareness of the topic in anyone who was not already acquainted with it, and an awareness of the catastrophic misconceptions which can ensue from too superficial an approach to it. I hope, in particular, to have undermined any preconception to the effect that rational courses of action are always optimal courses of action, and that to be guided by reason must always necessarily be best.

Why, however, is this preconception as widespread in the first place as I believe it is? What sustains it? Examples of self-defeating rational calculations are, after all, extremely common and widespread, and their basic features are not particularly difficult to understand.

Part of the answer probably lies in the prevalence amongst us of unduly individualistic modes of thought. We are very prone to treat society as an individual writ large, to use the accounts of the behaviour of isolated individuals which are so comfortable and familiar to us as metaphorical frameworks for our discourse about society. This, of course, immediately leads to a misplaced optimism about the role of rationality in society, since for the isolated individual the rational is indeed, practically invariably, the optimal. Another part of the answer no doubt lies in our inveterate respect for the power of reason. We are so accustomed to perceiving it as a liberating force in our lives that it has become very difficult to acknowledge that it can also be a source of constraint. If this is indeed a cause, then natural scientists are likely to be at least as vulnerable to its operation as any other group or occupation.

There is, however, probably another, less savoury factor which is also at work. The belief that people do best rationally to pursue their own individual self-interest has been an important component of political and economic ideologies for hundreds of years. It has been asserted over and over again in defence of competitive individualism, the principle of *laissez-faire,* and the continuation of a free market economy. And many people have wanted to establish the assertion so much that they have looked almost wholly to arguments and evidence in its support, and overlooked much of what counted against it. They have tended to take it for granted that rational egotistic behaviour is always to the advantage of individuals in society, even though this is palpably false. And they have gone on to seek for arguments which show that the individual rational pursuit of self-interest is also the best thing for society as a whole. To this end they have emphasized the role of competition and the good it brings. When people compete with each other in a free market economy it is their own self-interest which they seek. But their competition usually lowers prices, increases efficiency, maintains sensitivity to demand, makes for the most effective use of capital, and generally brings about a whole range of socially beneficial changes. Thus, since the eighteenth century, commentators have felt able to refer to an 'invisible hand' turning egotistically motivated actions to the common good, and to celebrate the free market system accordingly as the ideal mechanism for realizing mankind's collective aspirations.

Given this ideological tradition, the prisoners' dilemma has been rightly called 'the back of the invisible hand'.[6] It exemplifies all the awkward consequences of rational egotistical action in society, the consequences not denied but pushed into the background and made light of in much of the mainstream of our economic and political writing. Given that most of us draw a significant proportion of our ideas and images of society from this mainstream, an understanding of the prisoners' dilemma problem will serve us in good stead. It is a useful and versatile template for making sense of many specific social problems. And combined with what we generally know already it can produce a better and more comprehensive overall conception of the nature of society.

Note, however, that I have not so far said anything, in this section, about the *actual* nature of society. I have considered the consequences of rational self-interested calculations in social contexts, but I have not said how common such calculations actually are in society, as a matter of fact. Several important schools of thought believe that the tendency to rational calculation, and the concern to look to self-interest when making rational calculations, are natural tendencies in people, constituent parts of human nature. On this view people just are calculative and egotistical and there is nothing to be done about it. It is how we are, for good or ill. Our future will be determined by it, whether as inexorable progress to a rational society, or as nemesis, the catastrophic retribution which falls with tragic inevitability upon those who seek total, rational control of their own actions.

Both these alternative accounts of the future have their followings, although it is naturally the first, optimistic story which is the more generally attractive. Science plays a crucial part in this story. As the source of ever more and more reliable knowledge it is a progressive, liberating force. It makes people better and better informed, more and more able freely to calculate the consequences of their own actions over an ever-increasing range of conditions and in relation to an ever-longer time scale. This is the core vision of the most uncompromisingly optimistic accounts of the social role of science, the way that science appears in the most strongly scientistic forms of philosophy. Science is the cutting edge of a continuing process of rationalization. Scientific progress is leading to a utopia wherein human nature can be fully expressed, where all action is free individual action based upon self-interested rational calculation. And it is often thought that we are already very well along the road to this utopia, this particular scientistic version of the rational society.

There are, however, very good reasons for calling into question both the feasibility of this scientistic utopia and the conception of human nature which underpins it. First of all, there is the available empirical evidence. Even where they are readily able to do so, people do not devote themselves exlusively, or even predominantly, to rational calculation in the pursuit of their own individual ends. Even in the most advanced industrial societies, those most suffused with science and its idiom, vast areas of activity – perhaps I should say most areas of activity – defy interpretation as expressions of rational self-interest. One standard example is voting. No individual has ever made any difference to the result of a British election by going out to vote. Given that this is generally known, why does anyone ever take the trouble to do so?[7] Social-psychological laboratory experiments similarly indicate that the tendency to act on the basis of rational self-interest is by no means ubiquitous, even where no impediment is placed in the way of it. When people play prisoners' dilemma type games, for example, in experimental conditions, very substantial minorities, occasionally majorities, of participants do not choose the rational, interest-serving strategies.[8] And it is clear that this is not simply the result of error and mistaken calculation. The available experimental findings do not allow us to say precisely how self-serving and precisely how rational people in society actually are: there is an enormous can of worms associated with that question. But the experiments do call into question any suggestion that we are by nature wholly egotistical and wholly rational.

Finally, it is worth noting that the scientistic utopia provides what may prove *in principle* to be an inadequate account of a possible society. Any actual society must have tendencies to persistence and stability. Its own operation must, among other things, keep the society itself going. The society must be stable in the conditions created by its own operation. But nobody has yet succeeded in showing how a society based entirely and wholly upon rational self-interested calculation can meet these minimal requirements. And it is often argued that no such society can meet these requirements, that when rational self-interested calculation becomes ubiquitous in society, the basis of every single act, then there ceases to be a society left in which to calculate and the whole framework of calculation disintegrates. If this is true then the continuing existence of society must *necessarily* be based upon more than the self-interested reasoning of its constituent individuals, and it becomes a matter of relief to find that we are not, by nature, wholly rational and self-serving.

Appendix

This book is in no way critical of the trust we place in our current scientific knowledge: far from seeking to undermine that trust it shares in it. But it does again and again point to the dangers implicit in that trust. Our knowledge, it suggests, can be treacherous stuff; it must be handled with care and circumspection. It is fallible; it is not susceptible to conclusive and final validation; it is the magnificent but limited achievement of our own inventiveness. Few natural scientists would explicitly deny any of this. Yet to accept it, not as an abstract statement but as a direct description of the knowledge we all currently accept and use, is extremely difficult. The knowledge we routinely use, particularly the routine matter of fact empirical generalizations we routinely use, just seems right, a simple direct representation of nature, nothing more. At least this intuition can be very hard to resist and dispel.

To do this it helps to step outside our knowledge, as it were, and consider it as an outsider might. But this is something we rarely if ever do. We may look at the knowledge of other communities in this way, but not our own. By great good fortune, however, I am in a position to help in this task of detachment and external appraisal. I have in my possession the residue of a quite remarkable manuscript, in which a complete outsider offers a very critical appraisal of some of our own taken-for-granted everyday knowledge. Sadly, the original source of this material, 'The Flying Saucer Manuscript' as it was called, has completely disappeared in a regrettable incident which also involved the loss of the translation manual. But a fragment of the translation itself was saved, and I reproduce it below.

'PATERNITY AMONG THE BRITISH'

This note records some of the results of fieldwork carried out over many years among the British of north-west Europe. A general account of the institutions of that remarkable society will be given elsewhere. The present discussion is concerned entirely with their concept of 'paternity', a concept which they share with related cultures on the European mainland.

In brief, the British believe that the conception and birth of a child is a consequence of sexual intercourse between male and female. Moreover, a particular child is held always to be the consequence of a particular act of intercourse, and is said to be *related* to the particular male who performed the particular act. The male stands to the child, the British say, as 'father' to 'son' or as 'father' to 'daughter'. To establish which particular male was responsible for conception is to establish the 'paternity' of the 'son' or 'daughter'.

It must be emphasized that this extraordinary set of beliefs is uncritically accepted by the British. I never found anyone who questioned them during my stay. Establishing 'paternity' is a crucial institutional activity among the British, although mostly the matter is so 'obvious' as to need little investigation. The importance of 'paternity' is only apparent when its allocation is disputed. In such cases controversy and animosity of great intensity may be aroused, and the matter may be settled in a 'court'. Allocations of 'paternity' made by a 'judge' in a 'court' are generally accepted as definitive in British society. When a male is labelled a 'father' in such a context he must provide economic support for 'his' child under pain of imprisonment. For the most part however males routinely accept 'paternity' with respect to the progeny of their regular sexual partner or 'wife', and provide economic support for the 'wife's children' unthinkingly. The system of beliefs ensures, in a rough and ready way, that all progeny receive economic support, and in this sense operates quite effectively in British society.

Let us now move directly to the crucial question raised by this account. How is it possible for the British to maintain their faith in these beliefs? Why do they not, in the face of experience, recognize the futility of their ideas? They do, admittedly, recognize that, to conceive, a woman must be in touch with the gods (or more precisely the British recognize the sign of this – menstruation – without having any clear grasp of the basic idea). But they do not recognize the basic truth that conception for such women depends on the dice of the gods; rather, they see it as depending on an act of intercourse. They cannot see that their link between pleasant recreation and the procreation of their species is imaginary.

This truth is not recognized, it would seem for two reasons. First, the whole tissue of British thought is woven round their contrary idea. Let the reader consider any argument that would utterly demolish all British claims to be able to justify their viewpoint. Translated into British modes of thought it would in all likelihood merely serve to support their entire structure of belief. Second, the structure of British society is such that much crucial evidence against their views is practically impossible to locate and display. Much of the data which threaten to bring their belief system crashing into ruin is glossed over, or else surrounded by *taboos*.

THE IRREFUTABLE NATURE OF THE BELIEFS

The first reason is well illustrated by the British response to criticism. I was able to ascertain that most pair-bonded Britishers mate regularly and frequently, yet, needless to say, offspring were not at all regularly produced. But rather than admitting my claim that this established the independence of intercourse and conception (and suggested that the latter was random), the British treated this within their own framework. They did see the sense of my argument here, but they refused to admit its conclusions. A series of special excuses protected their own basic idea.

Many British for example believed that intercourse had to proceed at a particularly fortuitous time for conception to occur, and that this time could not be accurately predicted. This, of course, could readily be used as an 'explanation' of the randomness of conception against a background of frequent intercourse.

Even more remarkable was the widespread belief in the significance of 'semen', the fluid produced by the male at the completion of intercourse. The theory was that 'semen' is the initiating factor in conception, and that it works by somehow 'flowing up' the reproductive passages of the woman. This contradicts the normal British theory of fluids which are generally said only to flow downwards, but the British took little interest in this contradiction in their beliefs, and entirely refused to accept any need to change them.

Because of their 'semen' theory British often perform intercourse in ways designed to interfere with the normal ejection of the fluid, or perform rituals designed to destroy the fluid's 'potency'. Irregular birth times in the progency of regular mating pairs is often rationalized by the supposition that the pair have been 'spacing out' their family using these rituals. On the other hand, when the rituals fail to avert conception, as of course, they must, the British merely say that the rituals have 'gone wrong' and refuse to admit their futility.

Similarly, when women fail to procreate after many years of active intercourse, excuses are propounded. The male 'semen' may have lacked 'potency', or the female may have been a pathological specimen. At times completely *ad hoc* ideas are invoked, 'psychological tension' for example or 'fear of childbirth', may be interfering with the normal process. The British never see that these cases are evidence of the randomness of conception.

Conversely conception by virgins is held to be impossible, and when such cases occur they are not credited. 'Paternity' allocation procedures commence automatically. Indeed, such is the power of the 'paternity' beliefs that a virgin may be persuaded by her pregnancy that she must have had intercourse. In the face of the irrefutable 'evidence' of her pregnancy she may 'confess' to intercourse, and thus, actually count as evidence for the belief system of which originally she was such a blatant disconfirmation.

PROTECTIVE TABOOS

At several points during my stay in Britain I attempted to demonstrate the flaws in the native beliefs by practical demonstration. There was however a great reluctance to test their belief system on the part of the population. Females especially lacked any inclination to test their beliefs critically, fearing an unwanted pregnancy. Belief in the system sustained belief in the system.

But this was only one of a whole host of ways in which the beliefs were protected from empirical refutation. So many elaborate taboos surrounded sex that any open-minded appraisal of its relationship to conception was out of the question. Particularly awkward was the custom of indulging in intercourse only in strict privacy, which of course made any strict controlled demonstrations of the inadequacy of the local beliefs impossible. Again, it was impossible to compare samples who were having intercourse with a particular man with those who were having no intercourse – the norm of pair-bonding was too strong in society, and too strongly deterministic of permissible sexual relationships.

'SCIENCE'

For all the power of 'paternity' beliefs over the British mind, there was nonetheless some indication that rational criticism might have arisen in the society, but for one crucial institutional feature. At several points in my discourse with native informants I had the feeling that they were beginning to see the difficulties and inadequacies of their account. But rather than this resulting in a breakthrough of any kind the opposite occurred. I was told that on the particular point at issue I must consult a 'scientist'.

The role of 'scientist' in British society is needful of a whole book for its description. For now, it must suffice to say that they are the accredited knowledge manipulators. The authority of the 'scientists' backs the paternity beliefs current in the society. When a native is in difficulty justifying these beliefs he becomes conscious not of the inadequacy of his society's knowledge, but of the inadequacy of *his* knowledge relative to that of the 'scientists'. Hence rational criticism of beliefs current in Britain paradoxically leads to an increased respect for those who generate, disseminate and manipulate the beliefs – the 'scientists'.

I managed to talk to a 'scientist' about 'paternity' beliefs and found him in possession of far more sophisticated rationalizations than the typical native. He told an extraordinary elaborate story about 'semen'. Evidently scientists believe that it contains some sort of active principle, 'sperm', responsible for procreation. It is very difficult to convey his notion. In some sense it would seem the 'sperm' is thought of as living: in other ways it is regarded as perfectly natural and material. Certainly scientists seem to treat 'semen' just like any other fluid; it does not receive any special respect. Yet they seem reluctant to equate it with tears, urine, or any of the other routine bodily excretions. I remain puzzled about this peculiar concept.

Again, the scientific account of 'sperm' reveals curious paradoxes. Scientists have a corpuscular theory of it, and take the view that one single corpuscle suffices to 'fertilize' the female egg. Yet, at the same time, all males are believed to produce countless millions of sperm corpuscles at every ejaculation, and those males whose women do not conceive are sometimes told that it is because their 'sperm count' has dropped too low. Ejaculations of low 'sperm concentration', even though this is still of the order of countless millions of corpuscles per cc, are nonetheless regarded as certainly infertile. Once more, the evident contradiction was given no significance in British thought.

A FUNCTIONAL THEORY OF PATERNITY BELIEFS

It is clear that the beliefs described above cannot be described as accidental errors, or as beliefs about the world stemming from limited knowledge of it. I wish to suggest that those beliefs can only be understood in terms of the social functions they perform in British society. That is not to say that the British do not 'really' believe their 'paternity' beliefs, but rather that they really do believe them but only because their social functions have caused them to become accepted and transmitted from generation to generation on the basis of authority.

Two key themes must be perceived to understand the functions of 'paternity' beliefs. First there is the individualistic nature of British society and the very weak development of any sense of obligation to common property. Private possession, and the possibility of gaining personal prestige from possessed objects, are the institutionalized incentives for protecting and improving things. Secondly, there is the need to provide support for every child in the society.

Clearly the most effective means of ensuring this need in the light of British individualism is to provide specific links between particular children and particular adults with economic resources. As Britain is a 'money economy' and the male is the main money-earning unit, this is exactly how 'paternity' beliefs operate. The economic obligations of the British male to the young are channelled into the most acceptable form, given British individualism and materialism. Without this channelling system subjective feelings of obligation to the young would be much weaker. This is clear enough in practice from the relatively less good provision made by British society for young without living 'fathers'.

To conclude, a singularly beautiful and impelling piece of evidence can be given to support this functional theory. Over the last century in Britain (according to the excellent written records of the culture) the economic relationship of male and female has altered. The male is no longer the exclusive provider of economic resources, nor does he so strongly dictate the activities of his sexual partner. There is a move to sexual equality in the society. Imagine with what pleasure, then, I found a corresponding change in scientific theory. Over this same period the scientists have imposed on top of older paternity beliefs what they call a 'genetic' theory of inheritance. According to this, the 'make-up' of the child is determined by inherited elements called 'genes'. But the interesting thing about this recent idea is this, that exactly *half* the 'genes' in a child are held to come from the 'father' and half from the 'mother'! In the earlier dominant theories, on the other hand it was the male who contributed *all* of what mattered to the child. The realities of economic provision have determined the theories of 'genetic' provision!

submitted to the Martian Academy of Sciences . . .

Evidently, our erstwhile visitor from Mars held some distinctly strange beliefs, both about nature and about us. He could see no connection between sexual activity and procreation and treated them as quite independent phenomena. And since this was his belief about reality, he found us puzzling in that we believed something different. He was only able to resolve his perplexity by treating our paternity belief as a kind of expedient mythology, belief in which had fortunate results in our society. What he could not account for by reference to reality, he accounted for in terms of social 'functions'.

On the other hand, our Martian's beliefs about nature are perhaps not quite so strange as they seem. More than one human community has actually believed much the same thing: it is documented in the anthropological literature. And if his beliefs about nature are not so strange after all, then perhaps neither are his beliefs about us. Do we not ourselves tend to explain beliefs by their social functions when we think them false of reality? In our society is not Marxism sometimes thought of in this way, or conversely monetarism?

Clearly, it would have been a fascinating experience to converse with our Martian visitor. What a tragedy it was that his research grant apparently ran out, when he might have helped us to explore so many important empirical and philosophical issues. Still, perhaps what is now no longer available in reality might profitably be simulated in the imagination. The mind is a great laboratory for virtual experience: if we do our utmost to take his part we can construct a dialogue with our Martian in the mind.

There are a number of interesting thought experiments which can be carried out in this way, on the basis of the above material, but I want to suggest just two. First, imagine that you are able to meet and freely converse with the Martian in isolation, in a closed room as it were. You may allow the Martian to treat you as someone of unimpeachable honesty and integrity, and have him accept whatever of your personal experience and personal observation you put forward. But, in the task of seeking to press him towards your, or perhaps I should say 'our', point of view, you may not call in outside aid or place unquestioning reliance upon the opinions of authorities or other people generally: the Martian, remember, will be deeply sceptical of what you have to say, and very prone to respond to your claims or suggestions with, 'how do you know?'

The second thought experiment is simply a matter of mentally walking out of the closed room. Take the Martian to other people, to the authorities, to the great laboratories or the Public Records Office. Draw upon the whole of the resources of society to convince him he is wrong: find the best arguments, the clearest evidence, the wisest and subtlest advocates, and set them all to work. And think whether or not all our collective resources would force him into surrendering his views.

I must not guess as to the outcome of these thought experiments. And I confess that particularly with regard to the second I am not at all sure what the results will be. But I venture to hope that after the first experiment, any idea that the knowledge of individuals is directly sustained by their own experience may have been qualified to some extent. And after the second experiment the notion that knowledge is sutained collectively may perhaps prove just a little more plausible.

Notes and References

INTRODUCTION

1 Robert Oppenheimer, as quoted in W. O. Hagstrom, *The Scientific Community,* New York, Basic Books, 1965, p. 226.

CHAPTER 1

1 D. J. de Solla Price, *Little Science: Big Science,* New York, Columbia, 1963.
2 Among the writers whose views are so briefly alluded to here are Boris Hessen, Robert Merton, Joseph Needham, Thorstein Veblen and Edgar Zilsel. A good source of further reading on this and other historical issues, is W. F. Bynum, E. J. Browne and R. Porter (eds), *Dictionary of the History of Science,* London, Macmillan, 1981.
3 Further reading on what follows may be found in K. Pavitt and M. Worboys, *Science, Technology and the Modern Industrial State,* London, Butterworth, 1977.
4 C. P. Snow, *Public Affairs,* London, Macmillan, 1971.
5 Ibid., p. 19.
6 Ibid., pp. 35–6.
7 M. Polanyi, *Personal Knowledge,* London, Routledge and Kegan Paul, 1958. See also J. Ravetz, *Scientific Knowledge and its Social Problems,* Oxford, Oxford University Press, 1971.
8 T. S. Kuhn, *The Structure of Scientific Revolutions,* second edition, Chicago, Chicago University Press, 1970, p. 47. Also T. S. Kuhn, 'The Function of Dogma in Scientific Research', in A. C. Crombie (ed.), *Scientific Change,* London, Heinemann, 1963.
9 It is difficult to think of a good non-technical source to which to refer here, but see C. Norman, *The God that Limps: Science and Technology in the Eighties,* New York, Norton, 1981, especially chapter 3.
10 See, for example, Norman, *The God that Limps,* p. 73.

11 This is another important subject where it is difficult to suggest accessible further reading. But see *Daedalus,* 1975, vol. 104, no. 3, and particularly pp. 99–154; and also W. H. Armacost, *The Politics of Weapons Innovation: the Thor–Jupiter Controversy,* New York, Columbia University Press, 1969. A well chosen extract from this last work can be found in D. MacKenzie and J. Wajcman (eds), *The Social Shaping of Technology,* Milton Keynes, Open University Press, 1985.
12 D. MacKenzie, 'Nuclear War Planning', available from the author, Dept. of Sociology, University of Edinburgh.
13 M. Kaldor, *The Baroque Arsenal,* London, Andre Deutsch, 1981.

CHAPTER 2

1 L. H. Kern, H. L. Mirels and V. G. Hinslow, 'Scientists' Understanding of Propositional Logic: An Experimental Investigation', *Social Studies of Science,* 1983, vol. 13, no. 1, pp. 131–46.
2 Much of the material in the remainder of this section derives from the work of Robert Merton. See R. K. Merton, *The Sociology of Science,* Chicago, Chicago University Press, 1973.
3 S. Shapin, 'Pump and Circumstance: Robert Boyle's Literary Technology', *Social Studies of Science,* 1985, vol. 14, no. 4, pp. 488–520. Quotation on p. 488.
4 R. Westrum, 'Science and Social Intelligence about Anomalies: the Case of Meteorites', reprinted in H. Collins (ed.), *Sociology of Scientific Knowledge: A Sourcebook,* Bath, Bath University Press, 1982, pp. 185–217.
5 Ibid., p. 199.
6 Ibid., p. 194.
7 Ibid., p. 212.
8 Ibid., p. 191.
9 This is just one of a number of interesting episodes documented in B. Barber, 'Resistance by Scientists to Scientific Discovery', *Science,* 1961, no. 84, pp. 596–602. I have drawn further upon this valuable paper in what follows.
10 T. S. Kuhn, *Black Body Theory and the Quantum Discontinuity, 1894–1912,* Oxford, Clarendon Press, 1978.
11 Kuhn, *Black Body Theory,* chapter 8, note 10.
12 E. B. Mpemba and D. G. Osborne, 'Cool?', *Physics Education,* 1969, no. 4, pp. 172–5.
13 My basic reference for what follows here is J. Burchfield, *Lord Kelvin and the Age of the Earth,* London, Macmillan, 1975.

CHAPTER 3

1 S. Milgram, *Obedience to Authority,* New York, Harper & Row, 1974.
2 Steven Perrin and Christopher Spencer, 'The Asch Effect – a Child of its

Time?', *Bulletin of the British Psychological Society,* 1980, vol. 32, pp. 405–6.

3 H. J. Laski, 'The Dangers of Obedience', *Harper's Magazine,* 1929, vol. 159, pp. 1–10. Quoted in S. Milgram, 'Some Conditions of Obedience and Disobedience to Authority', *Human Relations,* 1965, vol. 18, p. 75.

4 As the basis for what follows I have used F. M. Turner, 'The Victorian Conflict between Science and Religion: a Professional Dimension', *Isis,* 1978, vol. 69, pp. 356–76.

5 T. H. Huxley, *Collected Essays,* vol. 2, London, Macmillan, 1894, p. 52. Quoted by Turner, 'The Victorian Conflict', p. 358, note 3.

6 F. M. Turner, 'Rainfall, Plagues and the Prince of Wales', *Journal of British Studies,* 1974, vol. 13, pp. 46–65.

7 Turner, 'The Victorian Conflict', p. 372, note 3.

8 F. Galton, *English Men of Science: Their Nature and Nurture,* London, Frank Cass, 1970, p. 260.

CHAPTER 4

1 For a more extended discussion of the concept of scientism with further reading, see I. Cameron and D. O. Edge, *Scientific Images and their Social Uses,* London, Butterworth, 1979.

2 The great French philosopher Michel Foucault offers many penetrating insights into expertise from this point of view. I only wish I understood him well enough to say something about him here. See M. Foucault, *Discipline and Punish: The Birth of the Prison,* Harmondsworth, Penguin, 1979.

3 See J. Habermas, *Toward a Rational Society,* London, Heinemann, 1971. The fifth essay in this book, 'The Scientization of Politics and Public Opinion', is particularly relevant. Besides the qualification made in the main text I must add that these essays date from the 1960s and that Habermas, an exceptionally dynamic and productive writer, would be unlikely to treat these topics in exactly the same way today.

4 D. Robbins and R. Johnston, 'The Role of Cognitive and Occupational Differentiation in Scientific Controversies', reprinted in H. Collins (ed.), *Sociology of Scientific Knowledge: A Source Book,* Bath, Bath University Press, 1982.

5 P. Raven-Hansen, *Technology Review,* May 1980, p. 34.

6 F. von Hippel, 'The Emperor's New Clothes – 1981', *Physics Today,* July 1981, pp. 34–41, quotation from p. 34.

CHAPTER 5

1 The best general survey of which I am aware concerning our relationship with technology and how it is likely to unfold occurs in a rather obscure place, but I note it nonetheless. It contains many useful further references. 'Technology and Social change', chapter 19 of R. Perruci, D. Knudsen, R.

Hamby, *Sociology: Basic Structures and Processes,* New York, W. C. Brown, 1977.

2 The discussion in this section is structured around a simple contrast of two opposed points of view: it suggests that whereas technology is often thought to control people, in fact people control technology. This is actually far too simple an approach, even as a very general guiding framework for thought. Consider for example how technology insinuates itself into the human imagination and flavours thought: here is a matter of enormous interest and importance which just does not fit within the present discussion at all. See for example J. David Bolter, *Turing's Man,* Chapel Hill, University of North Carolina, 1984, for one well-illustrated approach to this connection. For a much more philosophical and fundamental discussion of basically the same connection, and one heavily critical of modern 'technological rationality,' see H. Marcuse, *One-Dimensional Man,* New York, Beacon Press, 1964.

3 For sources offering properly thorough treatments of what follows in this and the next section, see B. Barry and R. Hardin (eds), *Rational Man: Irrational Society,* Beverly Hills, Sage, 1982.

4 For a proof of Arrow's Theorem see Barry and Hardin, *Rational Man.*

5 This statement is actually a gross over-simplification of the situation. I have said that a voting procedure may be exploited by its operators, but it may also be exploited by its voters. Anyone is at liberty to try to exploit a routinized practice and to try to use it to serve their interests. Whatever the system, one can try to think of a racket to beat it.

Is there a racket to beat the simple pairwise majority system and its operators? Perhaps there is. We can call it 'tactical voting'. Consider once more the awkward pattern which the operators of the vote may exploit.

Groups	*Rank order of preference*		
(all less than 50%)	*1st*	*2nd*	*3rd*
A	T	J	F
B	J	F	T
C	F	T	J

Imagine that the operators of the vote wish T to win. Accordingly they will leave T out of the first vote, and set J against F. The operators anticipate that J will beat F, because groups A and B will combine to defeat C; whereupon T will defeat J. since A and C will combine to defeat B. But voters themselves may perceive this as readily as operators. And in particular, voters in group B may perceive that they are doomed to their least favoured choice, T, by this sequence. Accordingly, group B voters may decide to support F on the first ballot against J. F thereupon wins the first ballot, and the second ballot, and group B gain their second preference instead of their third, which is advantageous to them. Instead of:

First vote J versus F
For J: (A + B) For F: (C)
J wins
Second vote J versus T
For J: (B) For T: (A + C)
T wins overall

we have:

First vote J versus F
For J: (A) For F: (B + C)
F wins
Second vote For F: (B + C) For T: (A)
F wins overall

Clearly, what I said earlier about the power possessed by the operators of the voting procedure is invalid just as it stands. By their tactical vote, group B produced a result contrary to that sought by the operators of the vote. Moreover, all reasonable persons in B would presumably vote tactically in this situation since it is in their interest to do so. So the operators are thwarted.

As it happens, however, the operators may turn out the victors after all. If it is general knowledge that group B are reasonable men, then the operators will know that the cause of T is lost if it is left to the last vote. But then it will no doubt also be general knowledge that A and C are groups of reasonable men too. So the operators will first of all set T against F. Groups B and C both prefer F to T. But group C, being reasonable, will vote tactically for T, the only option which can defeat the unthinkable J. Thus the operators will ensure the victory of their preference, T, by setting it *immediately* against the only alternative which is preferred to it, and not, as I incorrectly said earlier, by setting it aside from the voting for as long as possible.

But is this really the end of the story? Not so. Group B still has some hope. The group is about to be treated as so many reasonable persons, and thereby thwarted of its objectives. Therefore its members should pretend to be too stupid to vote tactically, or too ignorant of what the other groups are doing to work out tactics, or perhaps too attached to political principles to vote tactically. Any sign of stupidity, or ignorance, or principle, in group B might just persuade the operators to begin with J and F after all, whereupon the group could effect a coup by voting for F.

All these possible detailed calculations will be difficult to follow, and there are yet more which I could go on to point out. But such details are not important. What must be recognized is that everybody is having to calculate what everybody else is calculating, and in a situation like this there is no guaranteed best result. It is a salutary reminder of the limitations on the possibilities of rational calculation. Just as there is no best fixed method of playing poker, so there is no best fixed method of voting, or of taking the vote. The takers of the vote are not bound to have their way whatever strategy they arrive at; they have no determinate advantage over everybody else. But we can say in this case that most of the cards are in the hands of those operating the vote, and that their chances of prevailing stand high. The voting procedure, as it were, produces inequality of opportunity.

6 Barry and Hardin, *Rational Man.*

7 Note that some kind of a story can always be cooked up to make *any* action appear as the expression of rational self-interest. We could claim for example that people get a pleasurable sensation from voting, a thrill of self-satisfaction perhaps, or a feeling of self-importance, and that they go to vote to pick up the pleasurable sensation. Similarly, wherever we see something being done we can impute a pleasurable sensation to the doing, and thereby produce a consistent, if completely useless, general theory of human behaviour.

8 See for example R. Hardin, *Collective Action,* Baltimore, Johns Hopkins University Press, 1982.

Further Reading

This book has cut across a number of fields and discussed a very wide range of topics. If I were to attempt to give further reading for everything a truly immense list would result. Instead I shall suggest just a very few titles, nearly all closely related to the main themes of particular chapters, and mostly straightforward and accessible. Further references will be found in these few books.

CHAPTER 1

D. J. de Solla Price, *Little Science: Big Science* (New York, Columbia, 1963).
The seminal account of the exponential growth of science.

K. Pavitt and M. Worboys, *Science, Technology and the Modern Industrial State* (London, Butterworth, 1977).
A brief discussion of the context of the rise of science, with many useful further references.

D. S. Greenberg, *The Politics of American Science* (Harmondsworth, Penguin, 1969).
Fascinating material.

C. Norman, *The God that Limps: Science and Technology in the Eighties* (New York, Norton, 1981).
The most up-to-date general overview. Very useful.

CHAPTER 2

R. K. Merton, *The Sociology of Science* (Chicago, Chicago University Press, 1973).
This book includes the papers in which a systematic discussion of the scientific community and its system of recognition and reward was first presented.

W. O. Hagstrom, *The Scientific Community* (New York, Basic Books, 1975).

J. Ziman, *Public Knowledge: An Essay Concerning the Social Dimension of*

Science (London, Cambridge University Press, 1968).
Beautifully clear and accessible.

T. S. Kuhn, *The Structure of Scientific Revolutions* (Chicago, Chicago University Press, 1962).
A justly famous book. Every natural scientist should read it.

R. M. Pirsig, *Zen and the Art of Motorcycle Maintenance* (London, Bodley Head, 1974).

CHAPTER 3

S. Milgram, *Obedience to Authority* (New York, Harper & Row, 1974).

CHAPTER 4

D. Elliott and R. Elliott, *The Control of Technology* (London, Wykeham Publications, 1976).
A very clear discussion of technocracy.

D. Nelkin (ed.), *Controversy: Politics of Technical Decisions* (London, Sage, 1979).

D. Collingridge, *The Social Control of Technology* (London, Frances Pinter, 1980).

CHAPTER 5

B. Barry and R. Hardin (eds), *Rational Man: Irrational Society* (Beverly Hills, Sage, 1982).
Particularly useful for the editors' commentary and the further references.

A. Rapaport, *Strategy and Conscience* (New York, Harper & Row, 1964).
Idiosyncratic perhaps, but none the worse for that.

Finally, I shall mention two books for their general interest:

B. Barnes and D. O. Edge (eds), *Science in Context* (Milton Keynes, Open University Press, 1982).
This collection of article is not really suitable for independent perusal without any guidance. But it provides a general bibliography which points the way to many relevant bodies of further reading.

Thucydides, *History of the Peloponnesian War* (English translation, Harmondsworth, Penguin, 1954).
The relationship between science and politics is probably the most important topic discussed in my book. To think about it further it is necessary to reflect upon the nature of politics itself. This story of the career of a slightly over-vigorous democracy is definitely the place to start.

Index

Abel, N.H. 57
advocate-experts 108–9
age of the earth controversy 67–9
analogy *see* metaphor and analogy
answer analysis 108
anthropocentrism 8
anthropomorphism 8
Aristotelian cosmology 7
Arrow's impossibility theorem 129
artificial intelligence 90, 93
Asch, S.E. 79
Association of Scientific Workers 107
astrology 8, 20, 96
authority 72–9, *see also* scientific authority

Baroque technology 35
big science 4, 27
Biot, J.B. 54
British Society for Social Responsibility in Science 107
Boyle, R. 51
British Association for the Advancement of Science 56
bureaucratic politics 33–4

Capitalism
and the Scientific Revolution 14
Chemie Grünenthal 109
Chladni, E.E.F. 54
Clifford, W.K. 86
conformity 76, 79
creationism 94–5

Darwin, C.R. 57, 67
decisionism 99–100, 111
depoliticization 100
deterrent 34
divided expertise 105–10
division of intellectual labour 21, 57–8, 83

Education and training 16–19, 22–3
extra-sensory perception 63, 95

French Academy of Sciences 52–3
futurology 96

Galton, F. 86–8
Gauss, C.F. 57
global village 118–20
green revolution 118, 120

Habermas, J. 99–104
Hirohito, Emperor of Japan 56
Hooker, J. 86
Huxley, T.H. 85, 86

Industrial research laboratory 15
industrialization 1, 13–16
institutionalization 11–13
international competition 32, 117–18

Kelvin, Lord 68
Khan, G. 33

Laski, H.J. 80
Lavoisier, A.L. 53

lead in petrol 106–07
legitimation 97–8, 100–03
logistic curve 5
Lorentz, H.A. 57
Lyell, C. 67

Mendel, G. 57
metaphor and analogy 91–3
meteorite controversy 52–5
Milgram, S. 72–81, 88
military-industrial complex 26
money 44
Mpemba, E.B. 59
Mpemba effect 59–62
mutually assured destruction 31

Natural theology 84
new philosophy 8
Newton's law of cooling 59–60

Obedience 72–9
Osborne, D.G. 59

Parapsychology 63, 95–6
peer group pressure 76
phrenology 96
pluralism 112
Price, D.J. de S. 3–6, 10
prisoners' dilemma 134–7
problem of induction 78

Qualified scientists and engineers 4
quarks 65

Rational action cf. optimal
  action 132–4, 142
rational calculation
  individual and social 132–4
  in military context 137–8
  re scientific innovation 138–9
rationalist ideal of science 80–1
rationality in society 142–4
Rayleigh, Lord 56
research and development 28–30
  military 29–36
Royal Society of London 51, 56

Scholars and craftsmen 14
Science for People 107
science
  communication and quality control
    in 40–3
  cost of 4, 27–9, 140
  as craft 22–3
  critics of 23, 92
  and everyday understanding 20–1
  exponential growth of 2–5
  as form of cultural expression
    16–17
  and individual talent 37–9
  interdependence with other
    institutions 26–7
  as method 95–6
  origins of 7
  perception of 5–6
  and religion 83–8
  reward system of 44–9
  and technology 114–17
  utility of 17–18, 27–30, 114–17, 123,
    132
scientific authority 48, 70–1, 76,
  78–98, 110–11
scientific manpower 18–19
scientific naturalism 86
scientific profession 8–11
scientific reasoning 38–9
scientific recognition 45–9
scientific societies 8, 107
scientific training 10, 22–3, 40, 66,
  69–71
Scientific Revolution, The 7–8, 14, 51
scientism 91–4, 98, 114, 143–4
scientist
  first use of term 8–9
  cf. intellectual 9
self-interest 134–44
Snow, C.P. 18–20
social preference 124–32
Society for Freedom in Science 107
specialization 21–6, 57–8, 82–3, 88
spin off 29

Technocracy 19, 102–04, 110–12
technological determinism 119–21

teleology 8
toxicology 106–07
training *see* education and training
see also scientific training
Turner, F. 84–7
Tyndall, J. 86

UFOlogy 96

Vested interests 11, 32–5, 97
voting 127–31

Warnock Committee 122
Whewell, W. 9
witchcraft 8